W9-CII-926

TRAMMELL CROW, MASTER BUILDER

By the Same Author

The New Game on Wall Street
RCA
Salomon Brothers 1910–1985: Advancing to Leadership
IBM vs. Japan: The Struggle for the Future
Car Wars: The Battle for Global Supremacy
The Rise and Fall of the Conglomerate Kings
ITT: The Management of Opportunity
IBM: Colossus in Transition
The Worldly Economists
The Last Bull Market: Wall Street in the 1960s
The Fallen Colossus: The Great Crash of the Penn Central
Inside Wall Street
The Manipulators: America in the Media Age
They Satisfy: The Cigarette in American Life
NYSE: A History of the New York Stock Exchange
Herbert Hoover at the Onset of the Great Depression
The Entrepreneurs: Explorations Within the American Business Tradition
For Want of a Nail: If Burgoyne Had Won at Saratoga
Money Manias: Eras of Great Speculation in American History
Machines and Morality: The United States in the 1850s
Amex: A History of the American Stock Exchange
The Age of Giant Corporations: A Microeconomic History of the United States
Conquest and Conscience: The United States in the 1840s
The Curbstone Brokers: Origins of the American Stock Exchange
Panic on Wall Street: A History of America's Financial Disasters
The Great Bull Market: Wall Street in the 1920s
The Big Board: A History of the New York Stock Market
The Origins of Interventionism

TRAMMELL CROW, MASTER BUILDER

The Story of America's Largest Real Estate Empire

ROBERT SOBEL

JOHN WILEY & SONS

NEW YORK • CHICHESTER • BRISBANE • TORONTO • SINGAPORE

For Debbie D'Ambrogi

This publication is designed to provide accurate and authoritative information in regard to the subject matter covered. It is sold with the understanding that the publisher is not engaged in rendering legal, accounting, or other professional service. If legal advice or other expert assistance is required, the services of a competent professional person should be sought. *From a Declaration of Principles jointly adopted by a Committee of the American Bar Association and a Committee of Publishers.*

Library of Congress Cataloging in Publication Data:

Sobel, Robert, 1931 Feb. 19–
Trammell Crow, master builder : the story of America's largest real estate empire / Robert R. Sobel.
p. cm.
Bibliography: p.
ISBN 0-471-61326-6
1. Crow, Trammell. 2. Businessmen—United States—Biography. 3. Real estate development—United States—History. 4. Construction industry—United States—History. I. Title.
NC102.5.C79S63 1989
338.7'6908'0924—dc19 88-31170
[B] CIP

Printed in the United States of America

10 9 8 7 6 5 4 3 2 1

FOREWORD

H. Ross Perot

Gazing at the Dallas skyline, taking in the Anatole Hotel, Texas Commerce Bank Tower, and Trammell Crow Center, I cannot help but reflect that one man can really make a difference. The same thought occurs when I drive past that amazing complex of buildings that make up the Market Center, especially when I pass the Infomart. These are more than buildings constructed for one or another commercial purpose; they are tributes to Trammell Crow's genius.

I just wish everyone who picks up this book could actually see the structures that Trammell Crow has built. These are not the soulless slabs of glass and steel seen in too many of our cities, line after line of look-alikes that might have been pressed from some giant cookie cutter. Rather, each is different from the others, and yet they all reflect Trammell's exuberant, optimistic, and cheerful view of the world and its possibilities.

Over a period of several decades, his ability to dream creatively and make those dreams materialize marked Trammell Crow as one of the outstanding businessmen of this century. His fertile imagination has stretched far beyond Dallas, materializing in buildings sprinkled liberally throughout the United States and seven foreign countries, in giant warehouses and exotic hotels, skyscrapers and condominiums, farmland and strip malls.

Think about his creations for a moment and you will begin to understand something about this seemingly simple but actually very complicated man. There are many sides to Trammell Crow. We have known each other for many years, and each time we meet, whether for business or to socialize,

I recognize the essential, exuberant core of the man is always on display, but that there are delicate nuances as well, subtle shadings that are intriguing to observe. Trammell always has the knack of surprising even those who know him best. His is a portrait in primary colors, with pastel shadings. *Trammell Crow, Master Builder* captures both his optimism and complexity.

In this book is also revealed the man's imprint on the people of his remarkable organization. Truly a "self-made man," he can also boast of having made more of his partners millionaires than perhaps any businessman in our history. Some of the country's finest, smartest, brightest, and most ambitious men and women, with superb credentials, willingly start out ringing doorbells, seeking leases from clients, at a fraction of the remuneration they might receive in another industry. They do so because they have learned—not only from those with whom they interview but, more important, from the record itself—that if they make the grade, they will be employees for a very short time, that down the road a few years out is the promise of a partnership, a share of the action. Trammell Crow's people work harder and put in more hours than their counterparts elsewhere.

The hows, the whys, and the wherefores are set down clearly and entertainingly in this book. I won't give them away here, but I couldn't leave this matter without observing that Trammell's approach to cultivating people is, in retrospect, so simple, so obvious, and so successful that one can only wonder why other business leaders in other industries haven't used it as a model.

Finally, there is another important business lesson conveyed in this volume. Unlike many successful businessmen, Trammell Crow recognized, when the time came, that a changing business often requires changing talents to manage it. This in itself is a most remarkable and impressive achievement. In preparing for the long-term survival of his corporate empire, Trammell identified Don Williams as his heir and then stepped aside. He made certain that all in the company and those clients and associates who worked with him knew that he was now there as a consultant, mentor, and partner, and not as the leader. Trammell's widespread collectivity of interests is being transformed into a more durable structure that seeks to retain the spirit of his entrepreneurship while channeling human and capital assets in ways more in keeping with the altered business environment.

It is unfortunate that more people can't meet and converse with Trammell Crow. It would be an experience they would never forget, for he is one of the most charismatic men of our times. The best alternative is to read this book. This exciting story of a great business adventure offers clues for how entrepreneurs can succeed over the long term in our increasingly complex environment.

TABLE OF CONTENTS

INTRODUCTION

In order to understand the Trammell Crow Company, one must realize that for most of its existence it was not a sole entity or company. Even now there is some question as to whether it can be accurately designated as such.

On its face this is an odd statement. Certainly the same could not be said of any industrial firm, say, General Motors or International Business Machines (which are to their respective industries what Trammell Crow is to real estate development). Developer Samuel LeFrak would not be considered a company, but rather a sole proprietor engaged in a multitude of deals. On the other hand, Olympia & York, the large Canadian concern, certainly is a company, with a table of organization and all the other accoutrements one expects from such an enterprise. Trammell Crow Company is somewhere in between these two. It does have a structure—of sorts. But the men and women in the field, the developers operating under the Crow emblem, are partners or leasing agents hoping to become partners. They are not executives, and are jealous of their independence and prerogatives.

For most of its existence what will be called the Trammell Crow Company throughout this book was the reflection of Trammell Crow, one of those few individuals who can truly be called American originals. He did not create a new industry as much as introduce new ways of performing old tasks better. Nor did Crow originate management incentives, but rather devised a novel means of rewarding his associates. The same may be said for many of the hundreds of structures he erected all over the world—they are not new, but certainly many are different. For example, Crow certainly did not invent the modern exhibition hall and hotel, but his ideas about both drastically altered their shapes and functions.

While doing all this, Crow preferred to remain relatively anonymous. Still, he was always a significant presence, even when he finally recognized that his real estate empire required more formal organization than he could provide. When the scale and complexity of the business required different talents, Crow passed control to others whom he had helped train, and then stayed on to deal with special concerns and as a sort of consultant-expert and corporate symbol. In this too he was unusual; such relationships simply aren't supposed to happen or, if they do, work as well as this one has.

This is a firm with more paradoxes than most, which became clear on my first meeting with Trammell Crow. Not really understanding how the company operated, I innocently asked Crow if one day he intended to take it public. "We would be crazy to sell shares in what we have," he replied, "and anyone who bought them would be crazier still." Crow sits on the boards of 10 New York Stock Exchange–listed companies, but "Trammell Crow Inc." will never be one of them.

Later on I came to understand what he meant. It made little sense for Crow (and his partners) to incorporate and sell shares, because they could do far better financially by holding on to what they had. Moreover, it is vital to the success of the operation that all the principals be partners. In the pages that follow you will come to understand the Crow techniques, developed early in his career. He paid relatively low salaries but gave those selected as partners an equity interest in deals they were managing. If they succeeded, he would succeed. If they didn't, he wouldn't. It worked well. A few other companies may have created more millionaires than did Crow, but none of them did so as rapidly as he. In those early days, being singled out for partnership was a sure path to great wealth. It is the same today.

Trammell Crow came to real estate in a serendipitous fashion, and from the first ignored many of the rules others accepted as axiomatic, placing his unique stamp on individuals and operations. Much can be learned from his history, but there is no formula others can use, the kind peddled at real estate seminars or put forward in best-selling books. The reason is that Crow is one of the handful of authentic American business geniuses, a word not used lightly. He is a beguiling man who has always been able to fascinate and charm those he wanted to impress and instill in those who worked with him his business ideals and attitudes. Infused with the

Crow spirit, his associates were set off on their own, to work hard for themselves, but also for Crow.

Though there was a time when little more was needed to make the enterprise work, this is no longer possible. Today the Trammell Crow interests are so vast that newcomers have to settle for glimpses of the man and a knowledge of what he tried to accomplish and what they can perform and expect.

As one article on his techniques indicated, "Trammell Crow succeeds because *you* want him to succeed." If this motivation could be bottled and provided to other business leaders, the elixir would transform the American landscape. It can't, of course, so we will have to be content with what we have, the example of an extraordinarily complicated and sometimes contradictory man whose idea of sheer joy even now is to work a deal.

Crow is one of the greatest builders of our age. The Trammell Crow Company's warehouses alone account for more floor space than is utilized by the United States Postal Service. There are hotels, apartment complexes, office buildings, malls, hospitals, and display halls. The centerpiece of Crow's personal empire is the Dallas Market Center, a cluster of exhibition buildings in which retailers are shown products of related manufacturers, running the gamut from toys to men's furnishings, information processing equipment to jewelry. (The Market Center is not part of the company itself. It is owned by Crow's children.)

Adjacent to the Market Center is the Anatole Hotel, constructed, among other reasons, to house participants at marts and conventions held there. The Anatole occupies 48 acres of land. Managed by Loew's, the hotel advertises itself as a village in a city, which is pretty much what it is. Simply stated, there is nothing else like it. Uniqueness is a characteristic of Crow properties.

Crow had a hand in the architectural designs and in the selection of the art work at the Anatole, which is mostly Oriental, like that in the lobbies and meeting rooms of most of his office buildings and hotels. But Crow is not an architect and has never had formal training in art. He once modestly described himself to a visitor who complimented him on his taste in art as "an accumulator, not a collector." No one who views the collections on display in his offices, hotels, and other structures would agree.

In fact, Crow started out as an accountant. At the age of 33 he went into real estate development. Although even then he possessed the attributes that would take him far from a career in public accounting, little that he accomplished prior to 1948 hinted at the magnitude of his successes or the expansiveness of his tastes afterwards.

Despite his successes and talents, Crow never became a celebrity as, say, William Zeckendorf was in the 1950s or Donald Trump is today, to name two of the post–World War II period's more dazzling real estate tycoons. Indeed, when New York's Ed Koch was introduced to him in the early 1980s, the ebullient mayor seemed puzzled. He had heard of the company—in fact, he had asked it to participate in the renovation of 42nd Street. "I thought it was two people," he confessed. Yet the Crow holdings are several times bigger than Zeckendorf's and Trump's operations combined. When asked to compare Crow and Trump, one rival real estate tycoon snorted, "Donald Trump can look out of his window and see almost everything he owns. Trammell Crow is all over the world, and hasn't even seen many of the buildings he has erected and manages." Hyperbole to be sure, but the point is well made. As of mid-1988, Crow had constructed 225 million square feet of warehouse space, 60 million of offices, and 30 million of retail operations.

One of his creations was Trammell Crow Company, which appeared in a fashion that individuals involved with real estate will appreciate, but that others find puzzling. Until the mid-1970s—by which time the Trammell Crow interests were even then the largest on the American real estate scene—there really was no company. Rather, there were networks of common concern, scores of partnerships, small and large enterprises, all of which had one common denominator: the man. Trammell Crow.

How many partners did Crow have at that time? The answer can be found in the accompanying "Ownership Summary" chart, which was drawn in 1975. The Trammell Crow interests at that time resembled nothing less than a string of relationships, weblike in form, with Crow at the center. Note that he was not the only person with multiple tentacles—older, more senior partners likewise were encouraged to farm out deals to junior partners, and these in time would do the same with others.

The ownership chart was created to illustrate to Crow's lenders how complex the operations had become and to convince them to allow him time to work his way out of a serious financial situation. They did and he did, and in the aftermath of this crisis came the decision to fashion a structure for the holdings. The Trammell Crow interests were reborn. Or perhaps it would be more precise to say that safeguards were put into place, gingerly, tentatively, and only after the people who were involved were

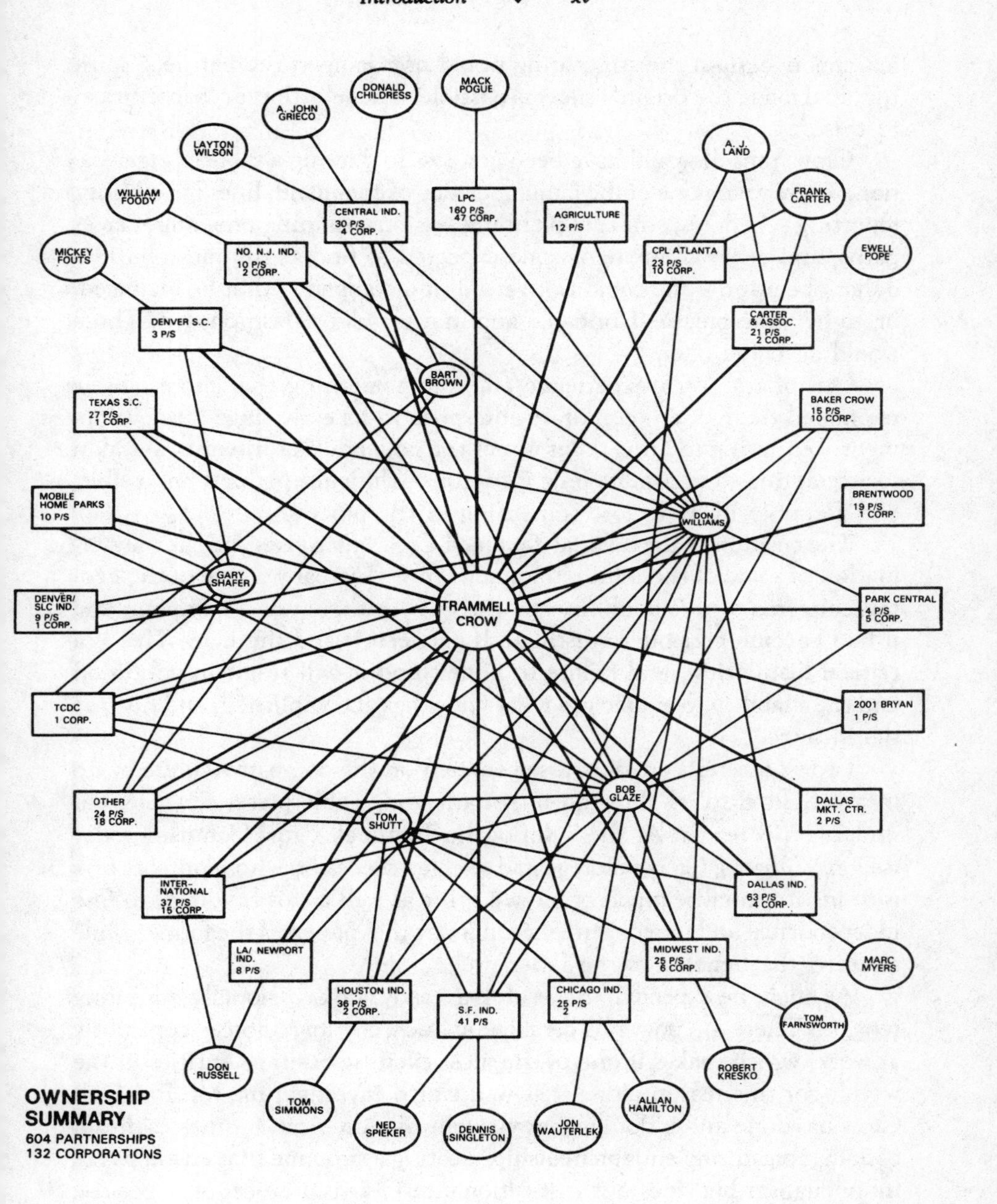

MACK POGUE
DONALD CHILDRESS
A. JOHN GRIECO
LAYTON WILSON
WILLIAM FOODY
MICKEY FOUTS
A. J. LAND
FRANK CARTER
EWELL POPE
LPC 160 P/S 47 CORP.
CENTRAL IND. 30 P/S 4 CORP.
AGRICULTURE 12 P/S
NO. N.J. IND. 10 P/S 2 CORP.
CPL ATLANTA 23 P/S 10 CORP.
DENVER S.C. 3 P/S
CARTER & ASSOC. 21 P/S 2 CORP.
BART BROWN
TEXAS S.C. 27 P/S 1 CORP.
BAKER CROW 15 P/S 10 CORP.
MOBILE HOME PARKS 10 P/S
BRENTWOOD 19 P/S 1 CORP.
DON WILLIAMS
GARY SHAFER
DENVER/ SLC IND. 9 P/S 1 CORP.
TRAMMELL CROW
PARK CENTRAL 4 P/S 5 CORP.
TCDC 1 CORP.
2001 BRYAN 1 P/S
BOB GLAZE
OTHER 24 P/S 18 CORP.
TOM SHUTT
DALLAS MKT. CTR. 2 P/S
INTER-NATIONAL 37 P/S 15 CORP.
DALLAS IND. 63 P/S 4 CORP.
LA/ NEWPORT IND. 8 P/S
MIDWEST IND. 25 P/S 6 CORP.
MARC MYERS
HOUSTON IND. 36 P/S 2 CORP.
S.F. IND. 41 P/S
CHICAGO IND. 25 P/S 2
TOM FARNSWORTH
DON RUSSELL
ROBERT KRESKO
TOM SIMMONS
NED SPIEKER
DENNIS SINGLETON
JON WAUTERLEK
ALLAN HAMILTON
OWNERSHIP SUMMARY
604 PARTNERSHIPS
132 CORPORATIONS

reasonably certain the structuring would not dampen the partners' spirit that had made the original success possible. The rebirth process continues to this day.

Crow himself would have been at a loss to draw up a table of organization. There were none of the familiar chains of command, lines of authority, reporting schedules, strategic planning sessions, training programs, career paths, and other components one expected to find at the multi-billion-dollar operation—one could not yet call it a company—that he managed, or, to be more precise, fabricated and in a very loose fashion, led. Those would all come later.

Out of the 1975 experience emerged something that more closely resembled the modern corporate enterprise in its early stages of development. But trying to bring order out of the collection of interests was akin to attempting to transmute jelly into stone—difficult, perhaps impossible, and certainly likely to lose something in the process.

The creation of the original Trammell Crow enterprises and the transformation of some of them into Trammell Crow Company is one of the central concerns of this book. By the early 1980s there was a company, and it had become possible to distinguish between it and the man. This is a critical distinction, and failure to understand it will result in confusion and the inability to appreciate what Crow has accomplished, and his true significance.

Crow's interests are of course centered on the company but are by no means limited to it. He is an important and visible presence there, but no longer its leader. At the helm of the Trammell Crow Company today is Don Williams, Crow's protégé and selected successor, whose primary task is to institutionalize much of Crow's spirit as well as his holdings, to fuse independence and interdependence in a way that preserves the former while realizing the benefits of the latter.

As might be expected, the result is a steady and occasionally frustrating tension. There are powerful centripetal and centrifugal forces constantly at work, which makes it one of the most exciting business entities in the world. For this reason alone it is well worth investigating, for Trammell Crow has done and is doing what virtually all American businesses dream of doing: organizing entrepreneurship, creating a structure that encompasses individualism but does not institutionalize it. It is a union of opposites: freedom and restraint. Once again, the fusion isn't supposed to work, but at Crow it usually does.

For readers who may be unfamiliar with the real estate industry, it should be noted that most developers do not seek to become "evergreen," a word often heard in Crow offices, which refers to the company's ability to survive the career or careers of the founders. Partnerships and corporations form only to fly apart, with all involved expecting just that. As will be seen, institutional survival is a key element at Crow. At the beginning of the 1980s it appeared to some that it too would not outlast the man. This much at least has changed.

It would be farfetched to think that so unique a company could provide a model for others, or that Crow is in any way a paradigm or symbol for American capitalism. But the man and the company do offer concepts other businesspeople and corporations might consider carefully, especially in areas of organization, flexibility, originality, and the forms that compensation can take. Indeed, almost any flourishing enterprise offers such possibilities. And Trammel Crow Company is one of the most successful of our time.

The reader will note a selected bibliography at the end of this book. It is understandably sparse; companies sometimes keep archives, but "non-companies" do not. Crow's papers consist of contracts, petitions, leases, and the like, and nothing else. Most real estate developers go from deal to deal and are unlikely to record or even reflect upon the wonders of the evolution of their business. In order to discover those facts that form the skeleton of any history, I have had to conduct interviews with present and past Crow associates, the Crow family, and others.

Fortunately, as a result of earlier efforts to create a history, I inherited more than a hundred interviews (and related manuscripts), most of them conducted by Andrew De Shong, a long-time friend and president of the Dallas Chamber of Commerce for many years, who was engaged for this purpose in the early 1980s. Over the years Trammell Crow has set down his ideas on a wide variety of subjects, in addition to reflections on his life and career, and he has been generous in sharing his recollections and reflections with me. His family, friends, and associates have done the same. David Christman, who teaches architectural history at Hofstra University, has made his expertise available to help me understand Crow's contributions in this area.

I have also had the unstinting assistance of several individuals at Crow headquarters in Dallas. Elsa Miller, Trammell Crow's personal assistant, shared her knowledge of the company with me and was of invaluable help in guiding me through the intricacies of the firm and the nature of the man. Betty Lane and Ann McGee, the departmental secretaries, provided information, assistance, and suggestions when they were needed. Bill Cooper, a long-time Crow associate and friend, shared his thoughts regarding Crow's methods of operation and the construction and management of the Market Center.

George David Smith of The Winthrop Group, Inc., not only contracted with the Trammell Crow Company for the book and brought me in as its writer, but also as a professional business historian provided much needed criticism of my ideas and plans, read and analyzed the manuscript in its various stages of preparation, and in all ways labored to make my work easier. Barbara Collins, Bob Kresko, Joel Peterson, and Don Williams functioned as a company review committee to assist me on matters of fact and interpretation, with the clear understanding that I was left free to come to my own conclusions.

My wife Carole accepted with grace my having to be at various parts of the country often at times inconvenient for her. Hofstra University permitted me release time to work on this project. Mary Curry of The Winthrop Group unearthed virtually everything in the public print about Crow, including articles in small, highly specialized journals and other more familiar works. Finally, Debbie D'Ambrogi acted as my liaison with Crow and was always there for consultation and to give requested—occasionally demanded—advice. She was a sounding board, resourceful researcher, indefatigable seeker of obscure information, and cheerful helper in a host of bewildering areas. All these individuals contributed to making working on this project so delightful. If the man and the company come alive in the pages that follow, they deserve much of the credit. If not, it is wholly my fault.

ONE

Dallas

I have to give Dallas a lot of credit for my success, while conceding it could have happened elsewhere. Maybe Atlanta would have done the same, maybe Los Angeles, maybe San Diego, maybe Denver. But it couldn't have worked in scores of other cities. What if I had been born in some out-of-the-way, remote place? How would things have turned out? The answer can never be known, of course, but it is the kind of question that others must ask themselves about their destinies.

—TRAMMELL CROW, 1987

One of Trammell Crow's favorite aphorisms is that while people can't do much about the cards life deals them, they can play them in a variety of ways. Another way of putting it might be that we are placed on Earth at a time, in a location, and in circumstances beyond our control; it is up to us what we make of opportunities that come along. Nine out of 10 people drift into what they do in life, and perhaps that is what happened to Trammell Crow in the beginning.

From what we know of him, it certainly couldn't be said that Crow intended all along to be in real estate. Yet there were present, in his boyhood and adolescence, forces that molded the man, determining his personality, values, outlook, and approach to life, and impelled him to seize the opportunity to enter that field when it was presented. The forces were complex and contradictory, as is the man himself.

For example, a number of years ago, in the early 1970s, Professor Howard Stevenson, who taught a course in real estate at the Harvard Business School, was seeking guest lecturers. Stevenson naturally turned to Crow, who was flattered to be asked. After the lecture, a student asked Crow to identify the most important single element in creating a successful business. Upon reflection, Crow replied, "Love." Crow firmly believes that if you truly have love for people you will be more successful with them. If you love what you are doing, you will be successful at your job. If you love yourself, you will be a better person.

Yet he also thinks that the best way to motivate workers is by offering them monetary rewards. Why has he been able to attract so many capable young people to his organization? Money. What keeps them there? Money. Why do they work so hard? Again, money.

Crow is a splendid example of cognitive dissonance. If it is indeed true that the mark of genius is the ability to hold two opposite views simultaneously, he surely qualifies as a prodigy. Crow sees no contradiction in the love/money dichotomy. His life was molded by love and a craving for success, which for him translates into money.

Trammell Crow is a complex, enigmatic person, as most of the great people of our time seem when analyzed carefully. Just as the secret of Orson Welles's Citizen Kane's life was wrapped up in "Rosebud," the sled he prized as a young boy, so Trammell Crow's personality can be explained by the seasoning he underwent while growing up in Texas between the early years of World War I and the outbreak of World War II. In a way, his experiences in this period were a metaphor for the generation that lived through those times.

Crow grew up in a large family. His paternal grandfather, William Jefferson Crow, fought on the Confederate side in the Civil War, and afterward became a cotton broker. William's wife, Sally Trammell, was originally from Alabama. They had seven children, one of whom was Jefferson, Trammell's father. According to Trammell, Jefferson was a small man—five feet, three inches tall—who as a result of a childhood injury did not have the full use of his left arm. He was also quiet, forceful, and devoted to his family.

Little is known of Jefferson's early life. He was born in 1874 in Longview, not far from the Louisiana border in northeast Texas, at the time a dusty town whose economy was based on subsistence agriculture and cotton farming. Jefferson was raised in Henderson, an even smaller town to the southwest. Jefferson graduated from high school, taught there for a while, and then was offered and accepted a job as a bookkeeper in nearby Monroe, Louisiana. It was there, at a Presbyterian church meeting, that he met Mary Simonton, the woman he was to marry.

Jefferson and Mary Crow were quite dissimilar. Mary was tall, beautiful, and wellborn; both of her grandfathers were physicians, as was her only brother. The couple's common interest was religion, which had brought them together. They married in 1907. Their first child, Brim, was born in 1909, and Kathleen arrived the following year.

In 1911 the family moved to Dallas, where Jefferson had found work as bookkeeper for Collett Munger, a local real estate developer who then was constructing a community in the northeastern part of the city known as Munger Place. As part of his remuneration Jefferson was permitted to live in one of Munger's houses, at 1318 Fitzhugh Street, on the edge of the new subdivision. The Crows would remain there until 1924. The dwelling consisted of a living room, one bedroom, a kitchen, and a sleeping porch. There was running water—via a pump in the kitchen—but no hot water. All but one of the other children were born at Fitzhugh Street—Davis in 1912, Stuart in 1913, Fred Trammell in 1914, Howard in 1916, Helen in 1918, and, seven years later, Virginia.

The Crows' lives centered around the church, the home, the workplace, and school. Theirs was a highly religious household. There was no work or play on the Sabbath, which began with services at the East Dallas Presbyterian Church at the corner of Swiss and Carroll. In the afternoon Jefferson read aloud to the children from the Bible or *The Youth's Companion,* and after dinner the Crows returned to church. There were prayer meetings on Wednesdays, too.

Jefferson Crow taught all his children to read before they started school and, as might be expected, concentrated on the Bible and other religious works. He also monitored the way they spoke. For example, the children were not permitted to use the word "bet." Rather, they had to say, "venture," as in "I venture he will be here in an hour." Trammell slides into such usages even now.

One might expect either tension or submission in such a crowded, strict, and constrained atmosphere, but apparently there was neither. The Crow children roamed the lower-middle-class neighborhood at will, perhaps because Mary tired of having so many of them under foot. The family would gather for the evening meal, and afterward would assemble in the living room to pray together on their knees.

Submission and modesty were virtues, the Crow children were taught. According to Davis, all the children except Trammell were quite shy and introverted. Although considered daring and outspoken by his siblings, Trammell was almost reticent when compared with other children at their school.

In 1924 the Crows purchased their own home at 6218 Prospect Street, in a development known as Empire Heights. After Fitzhugh Street it must have appeared quite luxurious; there were three bedrooms in addition to a sleeping porch, a dining room near the kitchen, a living room, and a bathroom with not only a lavatory and water closet but also a bathtub. The living room was heated by coal, and Mary cooked with kerosene.

The new home was a decided improvement, a sign of the progress Jefferson was making on his job. The fact remained, however, that they still had to husband scarce financial resources. All the Crow children recall the shabbiness of their clothes and the way they always seemed without money. Clothing was purchased during sales, the furniture at Prospect came used from Goodwill, food was simple and inexpensive, and nothing was wasted. At a time when some neighbors were buying cars, Jefferson rode a bicycle to work. The children were given lunch money and little else. Trammell remembers the nickel a week he received for Boy Scout dues.

Trammell and his brothers worked at odd jobs to earn pocket money. They mowed lawns, delivered newspapers, pressed clothes, caddied at a nearby golf course, clerked at grocery stores and filling stations, and unloaded boxcars at warehouses their cousin managed. Later on Trammell would frequently earn a dollar or two for driving a truck from the new Ford plant in Dallas to Fort Worth dealers' showrooms. As a result he was able to buy his own clothes and supply his own pocket money at a very young age. This was the expected mode of behavior. He didn't know it could be otherwise.

Dwight Eisenhower, who had been born in Denison, Texas, once remarked that his family was poor, but the wonder of America was that they didn't know it. This might have applied to the Crows, whose concerns and horizon were quite circumscribed. God and the church were real; material wealth, represented by new clothes, automobiles, vacations, and trips to the movies, was a chimera.

Jefferson performed volunteer work for *The Radio Revival,* a local gospel program featuring a Dr. Hawkins, who broadcast daily in a fashion still heard in that part of Texas. Perhaps that is why Jefferson indulged in one of the few extravagances of his life—the family purchased its first radio in 1928 or 1929. As far as frivolities were concerned, that was about it.

Jefferson Crow was not unambitious, and he wanted to provide well for his family. On the other hand, he was at peace with what he was and content with what he had. He worked hard and long and wanted to do well. Jefferson would go into the field with Munger, and on occasion took one of the boys along. He drove stakes into the ground to outline the lots and performed whatever other duties were required. From time to time he tried to sell real estate, and once came close to closing a deal. Trammell remembers those trips to the building sites and the euphoria at home when Jefferson believed he would receive a large commission for the sale—which fell through. Perhaps Jefferson's son filed away the experience and the feeling, drawing upon them later on. What he saw at the time, however, was a good and decent man destined to be a low-paid bookkeeper for the rest of his life, without ever having much in the way of material possessions.

Trammell did not become aware of his underprivileged economic circumstances until the sixth or seventh grade in school. His reaction was not one of depression as much as resentment and determination. Trammell did not blame his parents for the situation. Rather, he recalls harboring an anger for their lot as well as his own. He knew he did not want to live this way forever, and that no one but he could bring about a change. The idea obsessed him while still quite young. Trammell came to appreciate that whatever he was to get out of life had to be earned, that nothing would be given to him. If he was to rise in estate, Trammell would have to work hard and long. "All the responsibility for my life rested on me, all on myself, and solely on myself," he once remarked. "I could do things for others, but I couldn't expect others to do things for me."

Trammell respected and loved his parents and accepted their moral values. Yet he did not embrace his parents' path. The dilemma was finding a way to incorporate their values in the altogether different way of life he intended for himself.

Munger died in 1928, and now things became worse for the Crows. Jefferson lost his position. Like so many other Texans of his age, he would never again work at a full-time, steady job. The Texas economy had performed well during the 1920s, when cotton and petroleum led the way to a prosperity that many of the state's inhabitants shared, but none of this had affected Jefferson Crow. By the early 1930s he was one of the more than 20 percent of the workforce looking for a position of any kind, and unable to find one. He had a succession of part-time jobs: keeping the books for a small lighting company and a printing shop, for example. As was the case with millions of other families in similar predicaments, the fear and uncertainty affected the children, causing deep wounds in some, bitter determination in others, and scars that remain to this day.

Davis Crow left high school in the late 1920s and took a job at the Sun Oil Company; he was transferred away from Dallas, married, and established his own household. At about the same time, Stuart went to live with a maternal aunt in Mississippi, remaining with her until the mid-1930s. Brim, the oldest child, managed to win a scholarship to the University of Texas in Austin and moved there. Kathleen and Trammell remained at home, and both made contributions to the family finances. Kathleen found a position as a clerk in a doctor's office, receiving $10 a week, and Trammell worked before and after school, being paid $1.50 a day. Once he took a seven-day-a-week job as a soda jerk in a drug store, but had to quit when his father found out and would not permit him to work on the Sabbath.

Trammell was 16 by then and had already begun to resent what he considered his parents' unworldly attitude at a time when work was so hard to find. "I yielded out of obedience," he later recalled, "but my thoughts were my own. By this time I had come to feel that I was the family leader, and that my mother and father were naive and inadequate in many ways of the real world. I decided that they didn't get the whole picture, and I made up my mind then to be somebody."

Jefferson Crow would die in 1955 knowing that his son had achieved that objective, but the success meant little to him. When Trammell told him of his activities, Jefferson would listen and finally ask quietly, "But Trammell, how are you with the Lord?"

"Being somebody" did not include academic achievement. It was not that Trammell was a poor student, but academic matters bored him. He yearned for recognition, however, so he joined the high school's Reserve

Officers' Training Corps and tried out for the track and football teams. The R.O.T.C. didn't work out; Trammell resented taking orders from other students. He wasn't cut out for track either. Trammell did show some talent for football, though he was never truly interested in the game. He simply wanted to get an athletic letter before graduating, which he did.

Upon graduation in 1932 Trammell hoped to attend college, but under the circumstances knew he would be unable to do so without financial support. There was one way it could be accomplished: His family's dedication to religion led to an offer of a scholarship from the Presbyterian Church—if he would study for the ministry. By then Trammell had enough of organized religion, and he rejected the proposal. There was a chance for a football scholarship at Centenary College in Shreveport, but he failed to make the team. The University of Texas at Austin was about the least expensive school he might have attended; that was why Brim had gone there. For several weeks Trammell hitchhiked between Dallas and Austin, hoping to find a job that would enable him to work his way through college, but there was none to be had in what turned out to be the bottom year of the Great Depression.

So Trammell remained in Dallas, working at odd jobs for more than a year. He plucked chickens, cleaned old bricks for 15 cents an hour, and took whatever work came his way. Then, in 1933, he won a job as runner, at $13.50 a week, for the Mercantile National Bank in Dallas. He obtained this position by achieving the highest grade on a test the bank gave to a number of applicants. Now Trammell was able to further both his education and new career by attending evening classes at the American Institute of Banking, with the bank paying his tuition. Two years later, as a result of this effort, he was promoted to a teller's position.

By then Trammell also had a new goal: To become a certified public accountant. So in addition to the banking courses, he enrolled in accounting classes at Dallas College, the evening division of nearby Southern Methodist University. For the next three years he attended college five nights a week. Trammell was becoming more ambitious. For a while his plans included a law degree in addition to one in accounting, which would enable him to specialize in income tax management. So there were classes in law as well.

In 1938 Trammell passed the C.P.A. examination, one of 22 out of 420 who passed, and the youngest person to do so in Texas at that time.

He was now 24 years old, with growing self-confidence and a new optimism about his future. Things were looking up. He celebrated by purchasing a used car and, soon after, a partial interest in a small boat.

The following year, as World War II erupted overseas and the Depression was ending, Trammell found an entry level accounting post at Ernst & Ernst, where he audited books for local businesses. He left there in 1940 for a position with Smith, Morrison and Salois, then the city's leading accounting firm, where he was able to concentrate on income tax work. Trammell traveled all over the state, not only broadening his business horizons, but learning about other areas and people—especially people. He was considered an above-average tax accountant, but he was discovering that he was even better with people than with numbers.

In September, 1940, as the Battle of Britain raged and it seemed possible the Germans would soon invade that country, Congress passed and President Roosevelt signed the Burke-Wadsworth Selective Service Act. Trammell was now 26 years old, unmarried and in good health, and since he had a low draft number a likely draftee. He knew he could apply for a commission and utilize his background in accounting. As a C.P.A., he was offered a commission. Trammell opted for the Navy, and in April, 1941, as a new ensign, set out for Washington to report for duty in the Cost Inspection Division.

The next nine months were spent at the Bureau of Supplies and Accounts, the desk work supplemented by occasional trips to New York, Chicago, and Providence to audit the books of defense contractors. The work was fairly routine, but Trammell was seeing more of the country, gaining additional knowledge and experience, and thinking more seriously about his future after the war.

By then, too, he was contemplating marriage. The previous year he had met Margaret Doggett, the only child of E. B. and Lillian Doggett. E. B. Doggett was a prosperous Dallas merchant who with his partner, Homer Rogers, ran the Doggett Grain Company, a rather modest operation engaged in the kind of business fairly common in Dallas at the time—the purchase, storing, resale, and shipment of grain.

The Doggett family was quite religious, attending the nearby Methodist church regularly. Margaret's life to that point had been considerably different from Trammell's. As a child there were outings to the Doggett farm and trips to the Rocky Mountains in the summers, followed by attendance at Hockaday, an exclusive girls' school in Dallas, and then on to the University of Texas, from which she graduated with honors.

This untroubled existence came to an end with two dramatic incidents within months of one another. In March, 1939, driving back to Dallas with his wife after visiting Margaret at school, E. B. Doggett had a heart attack and lost control of his car, and both the Doggetts were killed in the subsequent crash. Homer Rogers took over at Doggett Grain, while a distant cousin, Miss Harlan Miller, became Margaret's guardian. That summer Margaret went to Europe with friends. She boarded the Cunard liner S.S. *Athenia* on September 3, 1939, for the return trip. War was declared that day, and that evening the *Athenia* was attacked and sunk, with the loss of 112 lives. Margaret spent several hours in a lifeboat before being rescued, and soon after came home on a highly publicized American neutral ship.

The following year Margaret and Trammell met at a party at the Peacock Terrace of the Baker Hotel, and they ran into one another several more times. Along the way they fell in love, and Trammell tried to wangle a transfer to Texas so he could be near Margaret. By then the United States had entered the war, and the naval shipbuilding program had been expanded. Destroyer escorts were being constructed at the Consolidated Shipbuilding Company in Orange, Texas. An officer was needed there to head the audit team, and Trammell got the appointment. As a result he and Margaret were married in August, 1942, and moved to Orange.

To those who knew them only casually it must have appeared a classic case of the girl from atop the hill marrying the boy from the other side of the tracks. Crow had to scrape for everything he got, did not have a college education, and were it not for the war might have been expected to continue on as an accountant; a partnership was the most he could hope for in that depressed part of the nation. Margaret was well-bred, reasonably wealthy, and educated. She was socially poised at that time, while Crow was ill at ease in groups of upper-class people, quite informal, and would always appear somewhat rumpled and preoccupied. He was then and always sensitive about Margaret's education and money and his lack of both.

This sensitivity spurred his ambitions and activities. Through reading, travels, and discussions he became an omnivorous devourer of ideas and theories. He was apparently interested in everything and everyone. In later years Crow would think little of telephoning his friend President Gerald Ford to ask about a political event about which he had heard or read, or of calling a professor who had written an article on a subject that was of concern. Put Crow into a room with an expert on any subject, and chances were that before the day was over he would have that person in a corner pumping him for information and ideas. Walking through one of his own exhibition halls years later, he came across a maintenance man sprinkling cleaning chemicals on the floor, and questioned him for half an hour about the process, how it worked, the costs, and what alternatives there might be to the materials he was using. Crow's desk was always piled high with books, and he often embarked on years-long reading programs in special areas. Perhaps this compensated for his lack of formal education. Decades later Crow would receive three honorary doctorates, of which he was most proud.

The same tenacity and dedication permeated his business activities. He was determined to prove to everyone, especially Margaret, that he could become wealthy on his own without using her money. The same man who claimed that the most important attribute of business was love could profess that the craving for financial success is the engine that inspires all effort. How much did Crow want? When asked this by a non-Texan who didn't know the local jokes, he would grin broadly, wink, and say, "I don't want all the land. I just want that next to mine." "If Trammell Crow could do it," said one close associate, "he would rebuild the entire world," adding that "it would be a better place for it."

When pressed, Crow will concede that the accumulation of a vast fortune is not all there is to life. Rather, as with many self-made businesspeople, wealth is a way to keep score, to measure one's success. Perhaps the drive to succeed is also a residue of the Depression, when his family was so poor, and of the time close to half a century ago when Crow felt the need to justify his worth to Margaret. But surely there is more to it than that. The man who once said that "work is more fun than fun" has always enjoyed what he was doing, finding in his occupations the kind of stimulation he craves. Crow is genuinely puzzled when close friends indicate a desire to retire and do nothing but play golf or fish.

Associates recall scores of instances in which Crow indicated his love of work. For example, an English businessman, visiting Dallas for the purpose of making some investments, was invited to dinner at Crow's home. Afterward the two of them and some of Crow's friends and colleagues sat around and talked of many things, including their aspirations and goals. "Trammell, I'd just like to get a little better view of your circumstances," the visitor said. "I've seen your library, I've met many of your friends, I see you're athletic, I understand your involvement in many affairs, and I'm quite impressed with all of the beautiful objects of art you have in your home. If you had an opportunity to add another dimension to your life, what would you want to do?" Without the slightest hesitation Crow responded, "I'd like to make another deal."

At Trammell's urging, Margaret later assumed the management of a chain of stores in Crow-owned hotels. Margaret had come to believe that there were better ways of operating such outlets, and this became one of her more important occupations. Today a large number of leasing agents in the Crow commercial operations are women, who win productivity awards disproportionate to their numbers.

Photographs of the Crows during the war show a couple who might have come straight from the pages of *Good Housekeeping* and *Life,* which featured articles dealing with the problems faced by returning veterans. Margaret was a diminutive woman with an open face and a warm and dazzling smile. Trammell was not quite six feet tall, already slightly balding, who in most pictures appeared the soul of earnestness. Acquaintances describing their physical impressions of him then remarked on his large, strong hands, bushy eyebrows, and intensity. This latter attribute showed up most in his eyes and smile. San Francisco newspaper columnist Herb Caen once wrote of him as a man with "foxy eyes and folksy ways." Crow possesses a piercing glance that transfixes those with whom he speaks, coupled with a broad smile and sharp, singular sense of humor. Then, as now, he had a knack with words and was capable of turning vivid phrases.

As a young man Crow made a strong and favorable impression on those with whom he came into contact, and his personal power would grow as he became older and more experienced. He displayed a transparent sincerity, which was coupled with unbridled optimism and certainty of what he was capable of and could accomplish, and a Messianic approach inherited from his parents. Many times he would discuss plans with associates, look them squarely in the eye, grasp their shoulders in those big hands, and paint pictures of what could be and how they would do it. Years later, journalists would write of "the treatment" employed by the then-Senator Lyndon Johnson in winning other legislators to his point of view. "He would envelop them in his arms and smother them with his personality," said a Johnson intimate. Johnson's fellow Texan, Trammell Crow, was also a master of the treatment. Both men were large, powerful, and energetic.

There was nothing calculated about Crow's expansiveness. He truly believes what he says and has confidence in his abilities. In his mid-70s, Crow took one of his many trips to China. Standing with some companions on the banks of a small river, with no way to cross, he stared at the other side, eyes narrowing as though concentrating on some great thought. Suddenly, without speaking, Crow waded into the water. He was soon up to his knees, and then, in the middle of the stream, to his waist. Did he know the depth of the river? Of course not, but Crow *believed* he could cross it. His friends, who had witnessed this kind of behavior before, called out to him, "Trammell, part the waters so we can get across," and one shouted, "I thought you were going to walk *on* the water, not under it."

In his prime Crow explained that such dramatic actions were not the result of overweening pride and unrelieved faith.

From some of my actions one might conclude that I'm not a person who has been plagued with self-doubts. When I make a decision, it's made. But it takes time, often a great deal of it. Frequently I'll keep something in my mind for months. Like a cow chews its cud, I'll keep bringing it back up. But then, when the decision is made, that's it. I've rarely been troubled with waking up at night wondering whether it was the right decision. I've never stopped a project once I've started it. But then, I don't enter into commitments frivolously.

Action is what counts. Too often businessmen are petrified by fears, await the unrolling of developments as though some inner logic will dictate the correct decision. I've never been impressed by such behavior. There is as much risk—even more—in doing nothing than doing something.

All this was to be in the future, of course, but the seeds for this confidence and approach had been planted during his youth and were nourished during the war. Trammell's sense of responsibility, self-possession, and knowledge grew along with his management of the materiel procurement program. Assignments took him all over the South and Southwest, and by 1944 he was a Commander in charge of cost inspection for the Eighth Naval District office in New Orleans. It was there Margaret gave birth to two sons, Robert and Howard, who were followed by Harlan, Trammell S., Lucy, and Stuart.

At the war's end in 1945 Trammell remained in the Navy for another year, frozen in place, to handle final settlements with naval contractors. In this period he gave some consideration to remaining in the Navy, but he abandoned the notion as impractical; he was too much of an individualist for a regimented existence. Besides, his experiences during the war had enlarged his ambitions. He was not certain what he wanted to do with his life, but he knew it would be difficult to go back to what he had been doing before entering the Navy. He still had not made up his mind when, in June, 1946, the Crow family returned to Dallas.

Fortuitously, Trammell had a near-perfect background for what was to be one of the major periods of expansion in American history. And there were few places that would benefit more from the growth than Dallas, Texas.

There were some fears that after the war, without the major government spending programs, the United States would fall back into depression. The conventional wisdom of most of America's leading economists held that the economy had revived as a result of war-related spending, the end of which would trigger a new slump.

The predicted disaster did not arrive; instead, the United States embarked on the most sustained boom it had ever known. In this period of expansion those willing to take chances would do far better than others who held back due to Depression-caused emotional scars. There had been a pent-up demand for goods and services because of the war, and there was also a large accumulation of savings available for investment. The raw statistics in the accompanying table indicate what was happening, but they cannot convey the exultant feelings that developed soon after Americans realized that the good times were not a mirage.

The years from 1945 to 1960 were characterized by steady economic growth, low unemployment, and, after the initial burst of postwar inflation, relatively stable prices. Construction activity was high as America experienced a real estate and building boom, especially in housing and commercial structures.

Texas not only shared in this growth, but in some ways led the process. The state's economy was based on three products: oil, beef, and cotton. From 1940 to 1948, the price of oil went from $1.19 to $2.25 a barrel, the beef cattle price from 17 to 43 cents a pound, and cotton from 10 to 30 cents a pound. By almost all statistical measures, Texas was among the top 10 states in terms of growth. Never before had the outlook seemed so unclouded, the prospects so pleasing. And Dallas seemed to be benefiting from the prosperity more than any other Texas city, with the possible exception of Houston.

Trammell Crow is a product of Dallas. The buoyant optimism he has always radiated is not peculiar to him, but can be found in key people throughout the city's history. A look at the development of Dallas's positive

SELECTED ECONOMIC STATISTICS 1939–1960
(figures in billions of dollars)

Year	*Gross National Product*	*New Construction*	*Unemployment Rate*	*Consumer Price Index*
1939	91.3	8.2	8.7	41.6
1945	213.4	5.8	1.9	53.9
1946	212.4	14.3	3.9	58.5
1947	235.2	20.0	3.9	66.9
1948	261.6	26.1	3.8	72.1
1949	260.4	26.7	5.9	71.4
1950	288.3	33.6	5.3	72.1
1951	333.4	35.4	3.3	77.8
1952	351.6	36.8	3.0	79.5
1953	371.6	39.1	2.9	80.1
1954	372.2	41.4	5.5	80.5
1955	405.9	46.5	4.4	80.2
1956	428.2	47.6	4.1	81.4
1957	451.0	49.1	4.3	84.3
1958	456.8	50.0	6.8	86.6
1959	495.8	55.4	5.5	87.3
1960	515.3	54.7	5.5	88.7

Source: *Historical Statistics*, pp. 135, 211; *1986 Economic Report of the President*, pp. 252, 314.

outlook illuminates the attitudes of her successful son—how Crow thought, why he acted as he did, and the reasons he was convinced his most ambitious expectations could be realized given hard work and concentration.

Dallasites today are aware that many people obtain their impressions of the Texas municipality from movie westerns and the television series *Dallas*. Perhaps viewers believe petroleum tycoons regularly stride through the downtown area in 10-gallon hats and spurs and that business is conducted on a predatory basis. While one occasionally sees those hats, business in Dallas is hardly like what television shows us.

In the first place, Dallas never was and is not currently a prototypical Texas city. It was not a frontier town; there were none of the serious troubles between settlers and Indians, cowboys and sheepmen, or other like material from which western legends are spun. Dallas lacked the natural resources base and port of Houston (the largest city in Texas and long-time rival), the tradition and sophistication of San Antonio, or the Spanish overlay of El Paso. It is one of the few Texas cities that might claim a French connection of some importance; in 1855 several hundred French men and women settled in a place they called La Réunion, and when the community failed they moved to Dallas. But even that flavor has disappeared.

Local history was not a preoccupation of Dallasites as they stood on the threshold of prosperity, however. Then—as now—they seemed more involved with the present and the future. Strolling through today's ultramodern downtown, they may not remember that prewar Dallas looked like a small town. A few undistinguished new downtown buildings stood alongside shabby wooden structures housing pawnshops and hot dog stands. Panhandlers were a common sight. Dallas in the years preceding World War II could be said to lack roots, class, and distinguishing characteristics. Physically, at least, it could have been any of a few score of similar-sized urban centers anywhere in the United States.

On the west side of Dallas the city limits paralleled the Trinity River, which is considered by some to be the divide between Central and East Texas, or, more expansively, between the American South and the West. The divide can be seen in the differences between Dallas and its neighbor, Fort Worth. Fort Worth is often called a "cow town" by some Dallasites, who in turn are viewed as arrogant outlanders by not only those from the neighboring city but by many other Texans as well. "Fort Worth is where the West begins—Dallas is where the East peters out" was a local truism. In the post-World War II period Fort Worth embraced the cow town characterization using "cowboy" terms in naming local businesses and promoting itself. Dallas accepted as its nickname "Big D."

Perhaps the pride of the "Big D" name is warranted by the fact that, more than most major cities, Dallas came into existence not so much because of geographic happenstance (such as accounted for the founding of New York, Chicago, and New Orleans, for example) but because of the efforts of individuals who arrived there for long forgotten reasons and of those who followed. Dallas was brought into prominence by the sheer willpower of her citizens.

There was no petroleum industry in the area until 1930, when Columbus M. (Dad) Joiner brought in an East Texas well (the Daisy Bradford No. 3) on a field from which more than 3 billion barrels would be pumped in the next quarter century. Firms extracting and refining the oil and servicing the industry set up their regional offices in Dallas; yet not a single barrel of oil has ever been extracted from Dallas county.

What was one of the greatest mineral finds in history was not an unalloyed blessing. Within eight months of the discovery, the price of oil skidded from $1.35 a barrel to as low as 5 cents. Nor did the bonanza mean much in the way of jobs for the locals. Roustabouts poured into Dallas from all over, and management was handled elsewhere. There was no work for the likes of Jefferson Crow and his sons.

The city had been a center for cotton brokerage long before 1930, however, and when Trammell Crow was a boy, Dallas housed the nation's largest spot cotton market. It did so by wresting leadership from other, more established areas.

Dallasites overcame their problems by enthusiastic effort. For example, through intense lobbying in the 1870s the city fathers made Dallas the preeminent rail head of the region, as the Houston & Texas Central came through in 1872. The railroad's management initially intended to bypass Dallas, but was convinced otherwise by $5,000 in cash, 115 acres of land, and a three-mile right of way. At about the same time, the Texas & Pacific Railroad was building from the East; the city's leaders were able to amend the state construction grant to require the line to pass through Dallas. When the railroad found a way around this requirement, Dallas offered funding plus a broad right of way, which altered perceptions somewhat. Dallas lost out to El Paso in winning the Atchison, Topeka, and Santa Fe, but not for want of trying. In 1881 the line's representative visited Dallas to arrange for a terminal, but an eager El Paso businessman happened to see him in a hotel room. Funds were offered and accepted, and so the Santa Fe went through El Paso. Not until 1955 was Dallas able to have constructed a spur from that line.

The railroads helped Dallas grow. In 1870 the town had a population of approximately 3,000, which expanded to 10,000 in 1880. Four years later, railroad tycoon Jay Gould observed that its location was as near ideal as could be imagined for economic growth and predicted that within a half century it would have a population of 250,000.* Other, even better-situated cities either never achieved their potential or declined after having done so, but Dallas was unstoppable.

The city attracted industry as aggressively as it did railroads. In Crow's youth it was one of the leading manufacturers of women's dresses, cotton

*The actual population in 1930 was 260,475.

gin machinery, and hats. It also was the preeminent fashion center for the Southwest, winning significant convention business.

Early in the century Dallasites helped pass laws to attract insurance companies to Texas, employing the familiar carrot and stick approach. The carrot was low capital requirements and relative freedom from regulation. The stick, fashioned in the state legislature by prominent Dallasites whose motto was "Keep the money in Texas," was a 1908 law requiring insurance companies doing business there to invest a minimum of 75 percent of their reserves in the state. Most Eastern companies left, new Texas-based ones took their places, and both groups relocated to Dallas. Similar incentives attracted the banking industry: low reserve requirements and a somnolent state regulatory commission. Dallas became the insurance-banking center for the Southwest, resulting in a large pool of money being available for businesses to draw upon.

It was this kind of effort that won for Dallas a Federal Reserve Bank in 1914, the year Crow was born. One story has it that on learning that a Cabinet member was heading to the Southwest to seek a location for the regional bank, several Dallas businessmen rushed to St. Louis where he was boarding a train and sold him on Dallas en route.

Dallas's modern history could be said to have begun in 1934, when Robert Thornton, chairman of the Mercantile National (the city's third-largest bank), attempted to lure the Texas Centennial Exposition to the city by promising $10 million in new facilities plus the use of 200 acres of fairgrounds. He was able to raise the funds by enlisting the aid of the city's most prominent bankers and other businesspeople. Houston also wanted the Exposition. A group of Houstonians published a small book entitled "Historic Places in and Around Dallas." All the pages were blank. Even so, this pledge, combined with some skillful lobbying, won the prize. The Centennial, with 13 million visitors coming to Dallas proved a great success, boosting the sagging economy.

Thornton next organized a group of 50 businessmen into the "Citizens' Council." It was later expanded to 200, and included the city's movers and shakers. "We didn't have time for no proxy people," Thornton later explained. "What we needed was men who could give you boss talk." Until his death in 1964, Thornton was a major factor in Dallas's growth, and the government-business nexus has never been broken.

This partnership is manifested in almost every aspect of the city's existence. For example, in 1945, when Chance-Vought was planning to relocate from Connecticut to Dallas, it was discovered that the airport runways at Love Field were too short to accommodate the company's planes. Unless something was done about this situation, the move would have to be canceled. Within less than three hours, the mayor and the city council met in emergency session and voted $250,000 for runway construction. In the 1950s, city and state political leaders lobbied successfully for prime participation in the Federal Highway Program, and as a result Dallas has one of the nation's most complete road networks and a major motor freight complex.

By the 1950s such accommodations seemed to come naturally. The aggressive "can-do" conviction motivated the creation of the Dallas-Fort Worth Airport, now the fourth busiest in the world and a facility unlikely to suffer from the congestion found in other municipalities, since its land mass is greater than that of all of Manhattan Island. If having a superb port was vital in the age of sea transport, an accessible airport in a region where weather poses no problem gives Dallas a decided advantage over other cities. Mayor Erik Jonsson observed, "The air is our ocean." Nor was that all; Dallas acquired a foreign trade zone, attractive to a host of businesses, and as a result half the air cargo in Texas comes through the Dallas-Fort Worth Airport.

When air conditioning made Dallas a more livable place, many service and "brainpower" industries were drawn to the city. The influx continued a tradition that had begun in 1874 when the newborn Dallas Board of Trade was formed by dealers in buffalo hides and dry goods, who announced their intention to transform Dallas into the region's commercial hub.

Dallas's population has been steadily—and sometimes spectacularly—increasing, as shown in the accompanying table. Remarking on the stream

GROWTH IN DALLAS, 1900–1950

Year	*Area (Square Miles)*	*Population*	*Increase*	*Percentage of Increase*
1900	9.78	42,638	4,571	12.0
1910	18.31	92,104	49,466	116.0
1920	23.54	158,976	66,872	72.6
1930	45.11	260,475	101,499	63.8
1940	45.59	294,734	34,259	13.2
1950	120.22	432,805	138,071	46.8

Source: *Dallas*, August, 1950, p. 20.

of newcomers in the 1940s, Dallas real estate promoter B. Hicks Majors said, "They began coming in from the North and from other parts of the South and West. These people had the energy of the Yankee and the thrift of the New Englander, and they took on the optimism of the Texan. But, most important, no matter where they came from, they right away picked up that great Dallas tradition of go-go-go." The enthusiasm was contagious from the start. Observing Dallasites in the post–Civil War period, a Fort Worth newspaperman offered that "The first thing the [Dallas] children are taught to speak is 'Hurrah for Dallas.' "

The Dallas scene has always been different from what one finds in New York, Chicago, Los Angeles, or other large American cities. Specifically, it is more freewheeling. Dallasites tend to be more willing to act on their informed hunches and are somewhat more trusting than their counterparts in other parts of the country. As Crow would later remark, "There are many more men in New York than in Dallas who could write a check for 10 million dollars—but there are probably fewer who would."

Much of this investment ardor derives from the optimism that seems to pervade Dallas. Local historian A. C. Greene once noted that the city is "a joyous place. It celebrates hope." He added: "You *must* be hopeful; admitting your business or your industry is weakening can be both economic and social suicide if you previously have made bold, encouraging predictions."

Another factor of Dallas's business climate was a decidedly negative view toward labor unions, an attitude that companies thinking about locating there took very seriously. An accommodating approach toward new enterprises was held by the city's leaders. Dallas has had a city manager system since 1931, and those who filled the post have usually been free from political taints and supportive of business. The city government has long been particularly friendly to real estate developers, passing less-restrictive zoning laws than in most urban areas its size. In fact, there were almost *no* zoning restrictions in the late 1940s, when under some circumstances builders could erect structures without even having permits. The lack of regulations enabled developers to build faster than their counterparts in those parts of the country where red tape held up projects for years, even decades. And the absence of strong unions meant lower costs and fewer delays.

Thornton, who did as much as anyone to make Dallas a world-class city, was mayor in the 1950s, and saw construction and prosperity as closely related. His mottos were: "I like congestion. It's better than recession,"

"You've got to build a city—the damned things don't grow like mushrooms," and "Keep th' dirt flyin'!" Dallasites did just that through the 1960s, 1970s, and into the 1980s. Starting in the early 1970s, there were years during which Dallas led the nation's cities in construction activity. In this period it was jokingly maintained that the official municipal bird was the construction crane.

Texas grew faster than most states during World War II, and afterward Dallas, with a population of more than 400,000 (having increased 25 percent during the war), vied with Houston for the title of the fastest-growing city in Texas. Houston remained the perennial population leader, but if one took into account entire metropolitan areas, Dallas could (and did) boast that it was the largest not only in Texas, but the entire Southwest; in fact, more than half the region's population in 1945 lived within 300 miles of downtown Dallas.

Growth was in the air. Building permits for 1946 came to a record $65 million. There were two Sears, Roebuck stores going up in the suburbs. Neiman-Marcus was adding to its old store and putting up a new one. General Motors Frigidaire Division announced a new regional office to serve 10 states. Certain-Teed erected its new asphalt roofing plant in Dallas. New hospital construction that year came to more than $15 million.

Warehousing in particular was prospering. Titche-Goettinger was completing work on a 100,000-square-foot warehouse, while McFadden-Miller, a rising force in construction, was busy putting up three block-long warehouses with more than 110,000 square feet. In August ground was broken for the first two buildings of the Alford Terminal Warehouses along Industrial Boulevard abutting the railroad sidings. Fred Alford, who dreamed of becoming the city's warehouse czar, spoke of the 17.5 acres of buildings he planned to erect, enough to handle 1,000 tons of merchandise a day and requiring at least 2,500 workers.

This flourishing postwar growth would be sustained for more than a decade before leveling off. By the early 1950s Dallas would have more warehouse space than any inland city—more in fact than entire states such as Florida and Oklahoma. By 1953 Texas ranked seventh in the nation in general merchandise space, and a quarter of the total was in Dallas. None of this success came as a surprise; indications that Dallas was in a fast growth mode were readily apparent in 1946 when Trammell Crow left the Navy and with Margaret and their sons moved into her family home at 3600 Armstrong Avenue.

As would thousands of veterans, Crow considered picking up where he had left off, which is to say returning to his old position at Smith, Morrison and Salois. He was received with seeming warmth there, and he knew the law required the accounting firm to take him on again. Yet Crow sensed a lack of enthusiasm on the part of his old associates, and he also felt he didn't really want to return to accounting. His wartime experience, marriage, and fatherhood had changed him from a person content with tax work and auditing the books of others to one wanting to control his own business.

For a few weeks Crow cast around for something to do. He had no difficulty perceiving the city's growth, and wondered how he might benefit from it. All those companies moving into the city had employees who would have to be relocated, which meant moving their household goods. Likewise, regional headquarters and new businesses would have to be furnished—more moving. So Crow became the Dallas agent for North American Van Lines, receiving the franchise for the operations in Dallas and New Orleans. In the process he learned about warehousing, since it is a vital part of that business. Yet the moving business didn't capture his imagination. Shortly thereafter Crow sold the franchise to his brother, Stuart, and went to work at the Doggett Grain Company.

Doggett was housed in a small, undistinguished, six-story, 70,000-square-foot concrete building at 425 South Field Street, which had been constructed in 1918. It had been designed in the days when cartage was just what the word says, conveyance by a cart, and when the space needs were relatively small and hand labor inexpensive. At the time, the company also owned grain elevators in Anton, Texline, and Littlefield, Texas—old wooden structures with some storage bins and a facility to unload grain trucks and load boxcars.

Doggett Grain's operations did not require the entire space in its Dallas building, and in fact most of it had been rented to Goodrich Tire, which used it as a warehouse, with Doggett occupying only a small part of the structure. During the Depression Goodrich had moved out, and at the time Crow arrived, most of the building was vacant, the only sizeable tenant being the Ray-O-Vac Battery Company, which occupied 4,000 square feet.

At the time, Doggett Grain was under the management of Homer Rogers, then in his late 50s. A likeable, intelligent, and decent man, Rogers eased Crow into operations and acted as his mentor. The two men sat across from one another, shared the work on an informal basis, listened in on each other's telephone calls, and continually exchanged views on a wide variety of agricultural developments and other matters. Initially, at least, Crow found the business fascinating and concluded he could have a pleasant if not outstanding career there.

Rogers and Crow made a good team, complementing each other nicely. The older man was by nature cautious, more often than not convinced that commodity prices would fall, while Crow was and is a congenital bull, who expected everything to perform well. Rogers didn't think much of the construction tumult, believing it was destined to end in a collapse, while Crow was becoming increasingly fascinated with all the activity, always talking about expanding the grain business.

Crow's instincts were right. Doggett tripled the size of its warehouses, erected new concrete buildings in Anton and Texline, and put up loading facilities in Leila Lake, Clayton, and in Spring, New Mexico. The company even rented an unused armament plant in Amarillo and stored grain there. Later on, however, as Rogers believed they would, grain prices did decline and the business with them.

In a different time Crow might have remained in the grain operation, but the kind of enterprise he and Rogers were running was on its way out. Increasingly, purchasers were going directly to the source, bypassing middlemen such as them. Shippers were sending their grain to port cities. The flurry of expansion in 1946, 1947, and 1948 was a last gasp. By the end of 1948 the writing was on the wall. Business was fading, and Doggett Grain's future was in doubt. Fortunately Crow was already considering a switch in careers. In fact, it already had begun, though at the time he wasn't aware of it.

TWO

Warehouses

> *If I can distill anything from my experiences it might be that it pays to go with the tide, and that fighting the generalized drift of history can prove a bumpy path. I wasn't about to do that, but at the same time, there was no master plan involved in the way I started out. Rather, it happened in a fairly simple fashion.*
>
> —TRAMMELL CROW, 1970

Sometime in 1948 Crow concluded that even had the grain business been better, it was not for him. The 33-year-old grain dealer decided to recast his career. Crow might have felt the need to prove himself capable of making it on his own. The realization that if he continued on at Doggett he would always stand in the shadow of the father-in-law he had never known might also have been a consideration, along with the suspicion that because of these shadows he would count for less in Margaret's eyes. Whatever the reason, Crow knew that he would have to try some other business in order to find fulfillment. That turned out to be real estate.

The raw materials for Crow's new occupation were all in place. He was an accountant who had specialized in income taxes and who knew how to assemble and interpret financial statistics. His work at the Mercantile had provided him with a fair acquaintance with banking and some contacts. The Navy training was a revelation, for during the war he had responsibilities usually assumed by men far his senior. He had learned management at the government's expense and had traveled from home for the first time in his life. Crow left the Navy with the awareness of having held positions and undergone experiences of a significant, maturing kind. Those five years in uniform, he later claimed, were worth a couple of M.B.A. programs.

Crow applied these lessons to the work at Doggett Grain, which as it happened was a near-perfect place to learn about the economy as a whole. Anything that occurs in the world affects the commodity markets, and to be a prudent grain dealer one needs to keep up with a wide variety of developments.

Finally, his employment at Doggett exposed Crow to architecture, though what he did hardly qualified him as an expert or even a journeyman. When he set out to build that new grain elevator in Texline, Crow and the manager decided to design it themselves in order to save money. There they were, two nonengineers, presuming to do the planning for the foundation, structure, walls, and the mechanical and electrical construction of a building going 30 feet into the ground and 125 feet above it. The undertaking was quite audacious and would not have been done as it was had Crow known as much about construction as he would a few years later. Nonetheless, the building went up, constructed by local labor hired mostly off the farms.

Naturally they made mistakes. Crow later realized that he probably put too much steel in the project and had made the foundations larger and stronger than they needed to be. The elevator is still standing, a monument to amateurism and unnecessary expenditure. Crow was finding out that there is nothing like doing something wrong to learn how to do it right. From then on he watched costs more carefully, never putting more into a building than required and could be returned in the form of rents.

One does not start out in real estate development in the same way as one begins a conventional manufacturing or service enterprise. There is no store, office, or factory to rent or inventories to accumulate. The would-be developer need not even be concerned initially with funding. What is required is a deal, and then another, and more to follow. Each deal is separate, and from each the developer obtains experience and, if successful, credibility.

In time the developer may reach the point of needing an organization, but that usually comes much later. Most developers—the ones who own a dozen houses or warehouses—can pretty well operate out of their homes, and even look upon their dealings as part-time jobs. This is how Crow began. While remaining at Doggett, he started to develop properties. Perhaps it would be more accurate to say that he drifted into real estate development rather than admitting, either to others or to himself, that a conscious change had been made.

The change began with that vacant space in the Doggett Building. Rogers had tried to rent it before Crow's arrival, but even in the tight market of the time few customers were found. That kind of multistory warehouse building was no longer suitable for many tenants, who preferred more accessible low-rises.

Crow decided to take a crack at leasing the space. He started out by contacting an old friend, John McFadden, whom he had met before the war. During the war McFadden had constructed housing quarters for Naval personnel, and had become an important force in warehouse construction. Crow asked him how to locate tenants for his building, and was told that the best people to talk to about such matters would be those at the Dallas Chamber of Commerce involved with industrial development. The Chamber of Commerce was Crow's next stop.

Some of the officials there gave Crow what turned out to be a good lead. Before the war, several floorcovering companies had jointly rented showrooms in the old Santa Fe Building but had been evicted because the space was needed by the Army's Eighth Service Command. They were at that time considering new showrooms, and Crow thought that he could fix up part of the Doggett Building for them. He discovered they were negotiating for space in a 10-story building not far from Doggett but had not come to any hard decision. So he set out to learn something about the showroom business, traveling to Chicago to wander through that city's famed Merchandise Mart and attend floorcovering shows. Then Crow returned to Dallas to see the floorcovering people, offering substantial improvements in the old warehouse, including a new elevator and other "amenities" he thought would serve them. He put himself in their place and tried to figure out what their needs might be. Crow also tailored lease terms and rents to their specifications. It worked. They struck a deal. This was his first experience in leasing, and it appeared he might have some talent for it.

McFadden and his partner, Edgar Miller, had put up a number of low-rise warehouses that had been rented by their owners to national concerns, and their firm seemed the logical choice to renovate the Doggett Building. Crow and McFadden shook hands on the deal, and they proceeded to make the changes. Such informal agreements became standard with Crow. In those days, whenever it was feasible, Crow operated on verbal understandings. McFadden & Miller would construct more than 12 million square feet of warehouse space for him in Dallas without ever having a written contract.

There was a rationale for this informal, person-to-person approach. Crow believed that the trappings of law are needed only where honor and trust are lacking, and that ultimately trust begets trust, with certainty and confidence being the result. With those old friends who were his first partners and associates, the approach seemed to work well. In dealing with a new associate, Crow would try to assess the newcomer's character, trusting his instincts in this, expecting the other party to do the same. Crow hoped the associate would find something good in him, and that he could live up to the expectations. Once there is mutual faith, he found, all else follows smoothly. Later on he would say that learning to depend on mutual trust was one of the first business lessons he learned in Dallas.

Before "charisma" became a buzzword, Crow was coming to realize that he possessed that elusive, powerful quality. Even as an auditor he had the knack of impressing people, and the talent was honed in the Navy and at Doggett Grain. The power of his personality flowered in the warehouse deals. Crow was able to command the loyalties and affections of a wide variety of people. He also perceived that trust and appreciation were good business, as well as a source of personal gratification. Certainly almost all who had dealings with him had no reason to regret giving him their trust and affection.

The floorcovering people moved in during the spring of 1948. Crow's initial exposure to real estate proved fulfilling, in addition to being financially rewarding, but he had some doubts about the matter.

While Crow did all he could to make the venture a success, from the first it seemed to him that exhibitors would be better served by more pleasant, open and airy locations. In those days wholesalers didn't give much thought to aesthetics. Their customers were retailers who supposedly would understand and accept the virtues of low-rent, rather undistinguished showplaces. Crow found this complacence questionable, and realized that in time he might want to provide a new type of exhibition space.

Shortly after the floorcovering wholesalers moved into the remodeled Doggett Building, Ray-O-Vac came in to say that its business was so good that additional space would be needed. Since there was none to be had at the Doggett Building, they would be moving when the lease expired. Earlier Crow might have let it go at that, but constructing the grain elevators had ignited his interest. Crow asked the Ray-O-Vac people where they intended to go, and on learning they hadn't yet made up their minds, offered to construct a new warehouse for them. He had some ideas at the time about what might be done if they accepted, but nothing was worked out in detail. Ray-O-Vac took him up on his offer, and he had to get to work in earnest.

Working in a way that satisfied his high self-expectations meant learning more about construction and financing and striving to be correct in every way, on every point. So Crow met regularly with John McFadden and others to find out all he could about investment possibilities, the problems and risks of warehouse construction, taxes, operating costs, and locating tenants.

Crow first visited the local loan office of the Equitable Life Assurance Society, and he talked with others he believed could provide information. He thought about and studied the matter from every possible angle. Crow later recalled having done 20 separate projections on the proposed Ray-O-Vac building. Everything seemed to add up—safe, foolproof, and even conservative. This seemed to be it for Crow—his big chance.

The Ray-O-Vac venture turned out to be the kind of deal frequently described at real estate seminars and in the now-ubiquitous books about making it big in real estate, but the concept was new to Crow and to most of his contemporaries.

The ideal situation was to have a property that threw off enough rental income to cover all payments and left something over for the owner. The next-best way was to structure a deal so the costs remained fairly constant while income rose due to inflation and renegotiated leases. At first a deal could have a "negative cash flow," which is to say the owner had out-of-pocket expenses even though the equity in the property was increasing, the hope being that soon the expense/income ratio would change. Finally, there were those deals in which the cash flow remained negative for quite a while.

All these property scenarios, even the last, could be profitable. For one thing, the tax laws enabled the owner to write off all interest payments, maintenance charges, *ad valorem* taxes, and depreciation. All the while, the owner was building up equity and paying off the mortgage. After the mortgage was paid, the owner would have a property free and clear and, given the nature of real estate, probably worth substantially more than the original price. In most cases the deal would be quite profitable, though in some the owner might have to wait several years before experiencing a positive cash flow.

Lacking adequate funds, one might find partners to put up money in return for a share of the ownership; the originator's contribution is to arrange the deal and carry it through. Crow thought about this option for quite a while, turning over the figures in his mind and working them out on paper, and he concluded that a partnership was not necessary. The deal seemed reasonable and "do-able" without outsiders. Of course, extending oneself in a deal is commonplace today, but not so in the late 1940s, when the scars of the Great Depression were still evident on the psyches of most businesspeople.

Although Crow started out with Ray-O-Vac in tow, it struck him that, given the expanding business climate, it was not necessary to have all the lessees signed up before proceeding with construction. This leap of faith implied a confidence in the future, which he surely possessed. It didn't make sense to him to require long-term leases either, for similar reasons. As he would later say, the auto company doesn't sell the car before it is assembled.

In those days, even in enthusiastic Dallas, it was the usual practice for builders of warehouses to line up tenants before seeking financing, or even a site. The goal was to sign them to long-term leases. Thus the builder could proceed, knowing the risks were limited and his financial future was as secure as possible. Much of what Crow did in this period seemed bold. Perhaps it was. Others did sometimes share his risk, however, and he always made a careful calculation of the risk/reward ratio before acting.

This kind of transaction is not unusual today, but then it seemed daring. Crow was considering erecting buildings for which there was no visible tenant—on speculation, as it were, at a time when speculation was a dirty word. Of course the decision to go ahead wasn't speculation at all, or even another instance of what has been called his perennial optimism. Rather, it was a realistic assessment of the situation. The city's real estate agents, having learned that he was renting better facilities for lower prices, were lining up to bring their clients to him. Not to have been optimistic in such circumstances would have been foolish.

Assuming that the location was right and the economy was moving along well, the building would bring higher rents the next time the leases were negotiated. Furthermore, even moderate inflation, say, 1/2 to 2 percent, would lead to higher rents, a quicker positive cash flow, and higher profits. In addition, Crow became convinced that long-term leases did not offer much in the way of protection. After all, companies could go bankrupt, leaving the landlord with no rent and no recourse, or they could come back and try to renegotiate a lower rent at a time when the owner is faced with overcapacity and a shortage of lessees. Finally, the two most competitive warehouse developers in Dallas were asking for long-term leases, leaving those tenants who wanted short-term leases to the others—meaning Trammell Crow.

The deal began with land, located in what Crow and many others realized was becoming one of the most important new sites in Dallas. This was the old Trinity River floodplain just west of the city, an area that might have been ideal for development were it not for fears that resulted from harrowing experiences with the river but that made no sense in the late 1940s.

The Trinity rises in north Texas, in the upland prairie south of the Red River, and then flows eastward toward Dallas, eventually winding up in the Gulf of Mexico. The 550-mile river drains an area the size of Connecticut. It was always a useless, unnavigable stream; historian Walter Prescott Webb called it "the rat-tailed Trinity," and an unknown commentator gave it the epithet: "that often unpredictable drainage ditch." When Will Rogers saw the river in the early 1920s and was asked his advice on what was to be done with it, his suggestion was, "Pave it."

When Dallas underwent drought, the Trinity would shrivel to a mere trickle, but when the rainfall was heavy, it could become a raging torrent, overflowing its banks and inundating the floodplain. When the settlers arrived, the Native Americans told them about a major flood that occurred around 1822, and the newcomers experienced similar overflows in 1844, 1866, 1871, and 1890.

There was a record flood in May, 1908. According to a local businessman, Leslie A. Stemmons, "The river went out of its banks and stretched two miles wide. A large part of the downtown was inundated. Eleven people were killed by the torrents, and 4,000 Dallasites were forced to flee their homes. Damage ran into the millions of dollars. If Dallas were to continue to grow, a solution to the problem had to be found."

One solution would be to create a system of dikes or dams, and another would be to alter the river's course. At the time the latter seemed more reasonable and less costly. Nothing was done due to recession and then World War I, but in the 1920s the City and County of Dallas Levee Improvement District went to work in earnest on a $23.9-million flood control program centered around the construction of a series of levees. More than 10,000 acres of land would eventually be reclaimed. The levees helped tame the Trinity; the floods were no longer as great a problem as before, though drainage difficulties remained. And there were financial difficulties too: In order to finance the work, in 1928 the Dallas Levee District issued $6.5 million in levee construction bonds and another $6.9 million for improvements on the Trinity River Flood Control Works.

By then the owners of land in the Trinity Basin, seeing the possibility of profits, decided to create a new company. A few weeks prior to the bond flotation they pooled their land and organized Industrial Properties, Inc., to be led by seven directors, all of whom were local businessmen. On July 21, 1928, the directors held their first meeting, at which one of them, J. D. Kervin, resigned, to be replaced by Stemmons, who soon became the driving force for the organization. Stemmons was one of the original supervisors for the district, as well as the leading land owner in the Trinity flood plain.

By the time the flood control program was completed, the nation was in the midst of the Great Depression, and construction was virtually halted. Most Industrial Properties members had lost heart, considering the land worthless, but Stemmons continued to see possibilities in the holding. He talked about transforming the floodplain into a thriving commercial center, an industrial park modeled after similar ones in Chicago, Kansas City, and Los Angeles. One day he took a group of railroad officials he hoped to interest in the property on a tour of the basin. As the conversation was later recalled, Stemmons told his prospects, standing at the intersection of today's Commerce Street and Industrial Boulevard, "Gentlemen, in 20 years this will be the busiest intersection of Dallas." One of the men was surprised. "Les, you don't mean that." But he did. The man shook his head and remarked, "My God, from cockleburs to congestion."

It was no idle dream. The Trinity Basin was quite close to the center of the city and to thoroughfares that led into Dallas. The Rock Island and the Southern Pacific railroads ran through it, and the Trinity was in good proximity to most of the motor freight houses. No zoning variances were needed. In fact, the basin wasn't even in the city limits at the time, and the county had no code at all regarding construction and occupancy, so potential builders were free to do whatever they could and would.

In a different period Leslie Stemmons might have worked wonders with the land, but not then. The bonds sold to finance the effort went into default, and until the arrears were cleared, nothing significant could be done in the basin. In addition to the bonded interest, an additional $1 million in back taxes was owed on the property.

During those years, Stemmons and then his sons John and Storey struggled to keep Industrial Properties intact. Their experiences made them prudent and restrained. Specifically, they had acquired a positive horror regarding debt, and they craved liquidity. Many years later, John Stemmons would adopt as his motto: "Never Put Your Name on Long-Term Paper." Yet the Stemmonses continued to have faith in the Trinity. Together they purchased the equity of Earl Harwell, a Tulsa oil millionaire, who after many years had become disgusted with the project. Other purchases followed, and in the end the Stemmons family emerged as the majority owners of the land.

In 1938, when there were some stirrings in the Dallas economy, the Stemmonses planned to put the company on a sound basis through another bond issue. When this failed, they tried to sell off lots to developers, hoping to use the proceeds not only to pay arrears, but also to attract developers to the area. They hired a New Orleans auctioneer and advertised as best they could, but few buyers came. Several lots sold for as low as $110, and these to speculators, not developers. Most of the lots didn't even draw bids. "We couldn't give them away," recalled John Stemmons 40 years later. "No one wanted them."

Leslie Stemmons died that year, and his sons took up the tasks of development. During World War II and immediately afterward they concentrated on three goals: to pay off the Levee District bonds, create a Dallas County Flood Control District, and then go on to sell lots. They succeeded in all three objectives. In 1944 and 1945 the Texas Congressional delegation managed to win authorization for an ambitious flood control program for the Trinity, which was funded in 1946. That same year the Levee District (meaning Stemmons) and the bondholders agreed to a restructuring of the debt, which effectively took the district and Industrial Properties out of distress. Under a new arrangement with the city, levee taxes on a typical lot would come to only $20 a year. Other taxes were reduced, through actions by the city and state, and the Army Corps of Engineers took a hand in the flood control situation. Now the Stemmonses went to work in earnest, putting in streets, arranging for utilities, and advertising for developers in what was a 10,000-acre park.

Industrial zoning came to the Trinity, and the Stemmonses had no objection to it; in fact, one of the first projects was a junkyard. Of particular interest was the fact that a large tract between the Texas & Pacific Railroad tracks and Turtle Creek was zoned for warehouses and light manufacturing plants of masonry construction. The Stemmons brothers knew they had a valuable property, and they assiduously sought builders and new business, advertising that "If you are looking for a new plant site, you—or your real estate dealer—are invited to call INDUSTRIAL PROPERTIES CORPORATION, developers of TRINITY INDUSTRIAL DISTRICT." They had a slogan to go along with the promotion: "Trinity Industrial District—Under the Skyline of Dallas."

The year 1946 saw the beginning of a construction boom in the Trinity, which meant the rejuvenation of Industrial Properties. Groundbreaking for the Alford Terminal Warehouses took place in August, 1947. The Texas & Pacific Railroad took an option on 40 acres for a $1.6-million terminal, and Continental Trailways announced it would erect a facility, to cost over $500,000. The Rock Island Railroad declared it would expand its freight terminal in the Trinity area. All these developments would make construction there—warehousing in particular—attractive. More than 90 sites had been purchased, with two-thirds of these either having existing buildings or having construction under way. The total expenditures for 1947 came to over $2 million; John Stemmons projected twice that figure for 1948. Most of the buildings were single-story masonry and steel warehouses, or else warehouse-like in appearance, running from around 12,000 to 80,000 square feet. They usually rented for from 40 to 80 cents per square foot per year, depending on the client's needs, location, amenities (in those days air conditioning was considered an amenity in these structures), and the like. Construction usually took 90 days, though in winter the time could run to 120 days.

Such was the situation in the old Trinity Basin when Crow and John Stemmons met. Those present at the time recall their meeting was at a Chamber of Commerce function, but Crow tells it otherwise. According to him, in the spring of 1948 Margaret and Trammell attended a flower show and were introduced to John and his wife, Ruth. It was the beginning of one of the most fruitful, and personally rewarding, business relationships Crow would ever know.

John Stemmons, with whom Crow conducted most of his dealings, looked like a "typical Texan." He was tall, rangy, and given to colorful and humorous expressions delivered in a regional accent. Stemmons could be smooth and charming when he wished to be. For years he had been one of the city's most powerful figures in politics and civic and cultural affairs. Now that the war was over, he was able to realize his father's ambition for the Trinity area, which was in the midst of a developing boom.

There were 15 buildings in the Trinity in 1947; by 1948 there were 45. In other words, Stemmons didn't need Crow; Crow needed him—specifically, for a small parcel of land as a site for the Ray-O-Vac project. They spoke by telephone the day after their meeting, and in a matter of hours Crow had purchased a 99-year lease on two lots on Cole Street in the Trinity.

The next step was to contact McFadden & Miller, who agreed to build the warehouse. Even though he was starting out on a thin financial thread, Crow decided to take chances. Ray-O-Vac contracted for 6,750 square feet of floor space, but the lot could accommodate a larger structure. So Crow opted for a warehouse with 11,250 square feet, leaving a vacant space of 4,500 square feet for which he would have to find another tenant.

Warehouses were then financed by getting a commitment from a long-term lender, who would offer the loan at an agreed-upon rate—in those days long-term rates were around 4½ percent—upon completion of the building and its occupancy by a stipulated tenant. Then the builder would take the commitment letter to a bank, which would lend him the money—called interim financing—to complete the project as required by the long-term lender. Next the long-term lender would buy the loan from the bank, and in this way the loan would be transferred from the bank to the long-term lender.

In practice, the owner tried to raise as much as he could up to and exceeding the cost. In those days, when Crow had no reputation, it wasn't a matter of choice, of course. He took what he could get.

Crow approached the Equitable for a loan, and soon learned it was unwilling to make one large enough to cover all his capital requirements. He then went to Pacific Mutual Life Insurance Company, which came close to covering the full price of the project. The next stop was the First National

Bank, where an old friend, Eugene MacElvaney, was executive vice-president. Crow took the loan application to the real estate department, which assembled all the necessary papers and agreed to make an interim loan once the permanent one was in place. Then the Pacific Mutual Life Insurance Company issued the long-term commitment. Crow took it to the First National, and then he returned to Pacific Mutual for another decision.

Pacific Mutual presented Crow with two loan payment options. He could have a 20-year mortgage with constant payments that would be fully amortizing, the kind that most homeowners are familiar with, or one on which he would pay 1/240th on the principal each month, plus interest. He elected the latter method, on the rationale that although the first month would be the toughest, afterward it would be downhill—gradually but always downhill—as smaller payouts were required each month.

Thus the financing was arranged, and the construction proceeded. The warehouse project was not large or complicated, but Crow gave it his close attention. He monitored the construction as though overseeing the erection of a skyscraper, watching the building rise from the ground, and visiting the site far more often than necessary.

The construction went smoothly, and Crow found a tenant—Decca Records—for the remaining space. The project was completed in three months, at a cost of $30,000. It generated a small, but positive, cash flow from the start. With this one successful deal under his belt, Crow felt sufficiently knowledgeable about the business to innovate. He planned another warehouse, but with a difference.

Even when refurbishing warehouse space for the floorcovering people at the Doggett Building, Crow had kept the tenants' concerns uppermost. It was reasonable that people would rather work in pleasant places with views of lawns, trees, and flowers than in stark environments with views of bleak vacant lots or other ugly structures. The ubiquitous old-style buildings, attached to one another in monotonous rows, with loading docks in the front and offices in the rear, were hardly appealing places for people to spend the workday. Crow felt that prospective tenants would prefer something better but somehow had never believed it possible.

So Crow set about to change people's image of what warehouses should be. For starters, the buildings would be freestanding. The loading docks and parking lots would be in the rear, with the executive offices in front, facing attractive landscaping. They would not even look like warehouses, but instead like office buildings, which in fact they were.

Without realizing it at the time, Crow was developing a principle that would be evident in most of his structures from then on, one that would set him apart from other developers of that period: In order to succeed, you must give the renter a better version of something he already knows he has to have. Crow began building warehouses in Dallas when there was a crying need for them. From the first, he built on the customers' requirements, filling them more than adequately. In the process, however, he recreated warehouses, and the customers found them preferable to all they had used before.

The opportunity for another venture presented itself much sooner than Crow thought it would. Some of the floorcovering tenants in the converted Doggett warehouse had the need for warehouses as well as showrooms and wished to combine both functions. One of them, James Lees & Sons, said it would require a one-story building that was more accessible, brighter, and less crowded. Clearly Lees and the others, including Karagheusian, Alexander Smith, and Bigelow, who were then Crow tenants, were prospective tenants for a suitable structure. They were interested and pleased when Crow offered to construct a building to suit their specific requirements. He had wet his feet with that first warehouse and was certain he could be equally successful with a multipurpose building of this kind, which after all would not be that much unlike a warehouse. Crow was convinced the Ray-O-Vac venture could be replicated again and again, and that once he developed a style, warehouses and display buildings could be turned out almost in cookie-cutter fashion.

Crow and James Lees & Sons drew up a lease arrangement for a building of 16,750 square feet, with a display area and space for offices. The lease would run for 10 years, with an option for a five-year renewal at the original rate with provisions to cover any tax increases. The annual rental charge was about 30 cents a square foot.

Crow knew there were enough differences between a pure warehouse and one with considerable showroom space so that this project would require the services of an architect. McFadden & Miller recommended one, Jake Anderson, who was employed at the firm of Harwood K. Smith. Anderson agreed to provide the drawings, working on weekends and at night.

Soon after that, Crow arranged a meeting between McFadden, Miller, and Anderson at Cole Street, where he had obtained an adjacent site from the Stemmonses on the same terms as before. From the first it was known that Crow had plans for much more than the floorcovering project. Along with the land for that building, the Stemmonses had given him an option for 585 linear feet of frontage along Cole Street, with a total of 73,125 square feet of land, and the Pacific Mutual Life understood that he would want additional loans. The frontages for his warehouses would vary—90 feet for the first warehouse, 135 for the second—depending on the needs of the renter. Crow thought he could erect four or five warehouses on that land, a forecast that at the time seemed quite wishful.

The four men who paced the Cole Street property were all aware of the ambitious plans. Crow marked off with his heel where he wanted the corners to be for the 12,500-square-foot building. He didn't have a tenant, but he proceeded anyway, calculating that the warehouse would be so attractive, this would not present an obstacle. He was right; soon after the work began, Kool-Vent Awning leased the property, and they moved into the building in January, 1950.

Other projects soon followed. Crow now had as much business as he could handle, and those 585 linear feet in the Trinity Basin were quickly gone.

Just when the work on the floorcovering displayers' project was completed, he found another company interested in a warehouse—Manhattan Shirts. Again Crow took a chance, putting up a larger building than was required, this one 25,000 square feet. Manhattan agreed to take 7,500 feet for a warehouse, sample room, and office. Before the structure was completed in the summer of 1950, Crow had signed on Mueller Brass, which leased 10,000 square feet for a warehouse. The Roney Company, a manufacturer of liquid petroleum gas equipment, took the remaining space for use as a warehouse/distributing center. Simultaneously, Crow erected a 15,000-square-foot warehouse at nearby Glass Street, which was leased before it was completed by Kellogg Co., a manufacturer of telephone equipment. Then came another 11,000-square-foot warehouse/office building on Glass Street, this one for the Brentwood Egg Company. By 1951 Crow had gone from managing a grain operation to being one of the most rapidly expanding builders in a city where there was no shortage of them.

There were others who shared Crow's visions and who had preceded Crow into the field. One of them was Fred Alford, who soon turned to other interests. Elmer Horn and Fred Wagner were factors for a few years; Horn simply left the business, while Wagner became increasingly involved with oil. Oilman Clint Murchison, who dabbled in many areas, acquired a million-dollar housing project and was negotiating for other properties.

Leo Corrigan was a very active force in Dallas (and other cities too) and was considered the most prolific builder in that part of the country. He put up several office buildings in downtown Dallas, as well as shopping centers and apartment buildings valued at over $100 million—but no warehouses. So while Crow had many competitors in this area of construction and renting, there was no dominant force.

There were several reasons for his early success. First, he worked hard, was careful about details, and brought in his projects on time and usually at or below cost. He visited his tenants on a monthly basis to see for himself that they were satisfied. He always tried to throw in something extra—a *lagniappe,* Creole merchants used to call it—not so much to make them feel they were getting a bargain, but more because he wanted to impress upon them the fact that he appreciated their business.

After a while the realization of Crow's ambitions became a matter of rote; he usually constructed far in advance of the paperwork—or even the blueprints. There were times when construction was begun before the deal had been finalized and the purchase of land was completed after the warehouse had been erected. Crow recalls a foreman asking him where a building was to be located, and after looking around the site, Crow walked to where he thought it should start and said, "Put a stake in the ground here, go 200 feet west, and then turn north."

Within a few months the Stemmons brothers realized how profitable those buildings were, and instead of selling land to Crow they contributed it for a part ownership in the structure. Crow and the Stemmonses created a different corporation for each venture, with each having a half interest. The Stemmonses would lease the needed land to the corporation, and Crow would hypothecate it to a mortgage company, which would provide the funds required for construction. This ownership pattern suited Crow, who by then had strained his relatively meager resources to the limit. It meant, too, that for the first time he had partners—in the parlance of the time, "outside partners," in that they provided land and left the business dealings to him.

Crow's contributions were time, effort, and knowledge; he was operating with other people's money and land. Was he taking risks? Certainly. But so were they. And his rewards would be greater than theirs when the projects proved successful. Such was the compensation for optimism.

Crow was keeping the dirt flying at the Trinity Industrial District, and he found himself having to make regular trips to the Stemmonses' offices to contract for additional acreage. "Now Trammell," John Stemmons once declared, "you say you want another piece of land. Where do you want it?" Crow looked at the map and started making marks—until he realized that he had asked for just about everything that remained. The ever cautious Stemmonses affectionately called Crow "our wild little brother," or, more simply, "the wild man." Their appellations aptly described Crow's point of view then and later. He was ever expansive, attempting to encompass all, reaching further than he might have thought prudent had he given the matter close consideration. Years later he would stand in a relatively undeveloped area with a partner and say, "I want you to buy all the land you can see from this spot." In time such individuals would learn that Crow did not mean this literally, but such utterances occasionally got him into trouble.

The Stemmonses learned about Crow's character early in the game. At that time there were many reasons in addition to Crow's land requests for them to celebrate. By early 1951 there were 155 structures either completed and occupied or under construction in the Trinity, and Crow was the largest builder in the district. Before they were finished, the partners together would construct more than 50 warehouses in the district, with approximately 2 million square feet of space.

By then Anderson had become Crow's full-time architect. Pacific Mutual Life handled most of Crow's mortgages, even though occasionally looking askance at his methods of doing business. Usually the lender came up with from 85 to 90 percent of what was needed to put up a warehouse, taking a lien on the property as collateral and leaving Crow to raise the rest. After a while, however, Pacific Mutual had loaned up to its limit. Crow knew he had to develop new sources of financing.

He went to several banks, and finally found that the First National in Dallas would furnish funds, but only if he put up some of Margaret's securities as collateral, which he did. So the embryonic developer coasted along at first, getting by on bank loans and trying to keep his equity interest in all projects at a minimum, conserving his limited capital so as to take on as many assignments as he could.

The financial coasting couldn't last for long. The projects were getting bigger and bigger. By late summer, Crow was engaged in constructing a $750,000, 127,500-square-foot master warehouse for B.F. Goodrich at Oaklawn Avenue and Turtle Creek Boulevard, for example, and was troubled about taking on so large a commitment on his own. He would need additional capital to expand as rapidly and securely as was possible. Crow looked to private investors for funds, once again turning to an old friend, this time Henri Bromberg, a lawyer whom he had known since the 1930s, when Bromberg's office was in the Mercantile National. Bromberg and his associate Edmund Kahn, both of whom dabbled in real estate, were most helpful. Bromberg did all the legal work, Kahn furnished the initial equity, and Crow located the sites and arranged for the financing and construction. They pooled their talents and energies, added a little money, borrowed a lot more, and put up warehouses.

Crow's arrangements with Bromberg and Kahn were as informal as those with the Stemmonses and McFadden & Miller. They collaborated on approximately 40 projects, primarily warehouses, over the next few years, without a written contract. As was the case with the Stemmonses, Bromberg and Kahn became close friends as well as business associates. In time Crow came to consider Kahn as one of his "brothers," a term he used to describe his dearest friends. Once he suffered a setback on one of their joint projects, and Bromberg and Kahn caught him attempting to shoulder all the charges himself without troubling them. "We had a hell of a time trying to get him to let us participate in taking the loss," Kahn recalled, "but that was the kind of partnership it was."

The business seemed to grow almost of its own accord. Crow and his associates developed a small warehouse empire in Dallas. Crow was on his way to becoming a local celebrity, at least in business circles. In 1951 he was asked to join the Chamber of Commerce and soon became active in its affairs. Significantly, he accepted a place on the committee to construct an east-west expressway for Dallas, which would open additional areas for real estate development.

All the while, Crow was breaking new ground. After the partnership with the Stemmonses he had the series of ventures with Bromberg and Kahn, who also became outside partners. Other, similar arrangements would follow, not only in Dallas, but elsewhere. These ventures were the beginning of what would become the Trammell Crow Company, but he truly didn't think about the arrangements in so structured a manner. Rather, Crow started out with a deal. Then there was another. A new partner came in for additional ones. Then more followed. He saw a collection of deals, not a company. In time some of those separate entities would be folded into the Trammell Crow Company, while in other instances Crow would retain interests in completely unrelated operations with several of the original partners. Bromberg and Kahn were not his only partners in Dallas. Crow constructed some warehouses there with Maurice Moore, an independent investor who also was chairman and principal stockholder in Continental Trailways, in which Crow had taken a small position. Crow had no clear idea of where this casual, but successful, process of cooperation was leading. But one thing was certain. He definitely was not thinking or dreaming of heading a large organization.

By 1953 Crow had put together what he called "Trinity Industrial Park," which was advertised by "Trammell Crow—Owner and Developer." Actually the "Park" was part of the Industrial District, but Crow labelled it separately to give it differentiation because of its more restrictive and higher development standards. The operation was grinding out warehouses for Talon, Clampitt Paper, Snap-Shots, Minneapolis-Honeywell, and other local and national companies.

Later in the year Crow tried his hand at something new—a building for the Trinity Industrial Bank, which as the name indicates was in the Trinity Industrial District to serve the growing businesses there. Crow not only helped design the structure, but also agreed to serve as a director of the bank. The building had several features unusual for the time, such as drive-in windows, specially reserved parking places for handicapped people, and no bars or windows in front of tellers' cages. By then, five years into the business, Crow was eager to try out his ideas about construction and services.

The inevitable happened—news of Crow's projects spread, often from companies which were pleased with their first warehouse and wanted others. Mueller Brass was one of these. The company's space in the Trinity was satisfactory, and after a while, in late 1950, Mueller told Crow it wanted another one—not in Dallas, but in Denver. So Crow flew to Denver, found a site, and constructed a warehouse there. Then Mueller wanted another warehouse in Atlanta. Mueller used a local agent, Frank Carter, to locate the site, and then brought Crow into the deal to put up the building. In a relatively short time Crow's horizon had broadened beyond Dallas and even Texas, to areas in which he had no ties. He found himself invited by landowners and real estate brokers to begin operations in other cities.

Crow realized there was no reason why he couldn't function throughout the whole United States, if only he could hit upon a proper method to do so without taking undue risks.

By the end of the 1950s, Crow had several million square feet of warehouses and could justifiably claim to be the biggest operator in that field. It seemed that he could shake his sleeve and a warehouse would tumble out. The man who had agonized so about his first project now directed construction on warehouses he would never see. Yet there is far more to real estate than warehouses, which Crow was coming to understand as he erected them on a wholesale basis in this period.

THREE

The Dallas Market Center

The idea is to keep costs down, and one way to do it is to borrow to the hilt for the longest time period possible. Preferable to a 20-year loan is one that runs for 10 years, and a 30-year loan to a 20-year one. The best kind, of course, would be one you never have to pay back.

—A TEXAS APHORISM

Warehouses have always been the foundation of Trammell Crow Company, literally as well as figuratively. They churn out earnings like clockwork, are easily maintained, and most of the time are little trouble. Like all foundations, they are hardly visible; few Americans are aware of them, and those who do notice them seldom understand their function or consider what their importance might be.

Crow appreciated the virtues of warehouses and realized he might easily continue erecting and managing them, thus enjoying a rewarding career. After a while, however, he set out in a different, though initially related, direction.

The next step in the evolution of Crow's vocation was into marts—large, wholesale marketplaces. The decision to take this path was a natural progression; just as exhibition buildings emerged from warehouses, so marts developed out of Crow's interest in exhibition buildings.

As mentioned earlier, Crow went to Chicago in 1948 to explore the Merchandise Mart before making his presentation to the Dallas carpet people. It wasn't the first time he had seen the mart. While in the Navy in 1942 Crow had been stationed in Chicago for a brief time, and one Sunday afternoon, as might any curious tourist, he visited the place. It was a fabled building even then. With its hundreds of showrooms under a single roof, the Merchandise Mart was a department store for retailers, where they could find everything their stores required at a wide variety of prices. Crow instantly recognized how well the mart's form served its function. He walked through many of its 24 floors, and he thought about the place often in the years that followed.

At the time, however, Crow entertained no ambitions about entering the mart business. Rather, he had become interested in a concept he thought both important and natural, a sensible way of displaying merchandise.

Consumers go shopping for all kinds of items, from shoes to hairpins to computers, without wondering how the goods they purchase come to be in the stores they patronize. How does the storekeeper decide what to stock? The answer is that the merchant goes shopping at trade shows, exhibition halls, or, if a large retailer has serious shopping to do, at a complex like the Merchandise Mart that had impressed Trammell Crow.

A while later, curious about how the mart idea originated, Crow did some reading on the history of such expositions and showrooms. They could be traced back to medieval regional fairs held at specified times of the year at such places as London, Reims, Champagne, Bruges, and Frankfort. The more important of these, such as those held in Champagne in January, lasted around six weeks and attracted merchants from all over Europe and even beyond. By the 18th century, permanent wholesale markets had appeared in London, Paris, Milan, and several other European cities.

In America in the 19th century, permanent wholesale fairs were established in New York, Boston, Chicago, and St. Louis. It was no coincidence that these mart metropolises were all port cities; merchants had to convey their purchases to their stores, and the cheapest way to do so was by water. The railroads transformed the situation, but tradition dictated that these four cities would remain the nation's preeminent wholesale centers, eclipsing smaller, regional ones.

Some regional markets managed to survive. The furniture industry provides an example. In Grand Rapids, Michigan, there was a small furniture mart, concentrating on upper-priced lines, that declined when the American Furniture Mart, concentrating on medium-priced lines, was constructed in Chicago after World War I. High Point, North Carolina, was a major furniture mart center, specializing in volume items. San Francisco and Atlanta had smaller, regional centers.

Before World War II the furniture marts held shows four times a year. These were expensive affairs because the bulky furniture had to be carted from factories to marts and then back again. (Such logistics difficulties were in part responsible for the persistence of the regional marts.) During the war, because of economic reasons as well as the difficulties of obtaining transportation, Chicago and Grand Rapids dropped their spring and fall shows. Major retailers, knowing that merchandise purchased in the summer and winter would arrive too late for their January and August sales, started concentrating their attention on High Point. In the early 1950s, then, the furniture industry was in need of small, regional markets.

Dallas had aspirations to expand into the wholesale exposition field. This wasn't a new idea. As early as 1859, there had been county fairs in Dallas, and in 1876, 17 Dallasites contributed $500 apiece to start a North Texas Fair. Nine years later a Fair Association was organized, and there were additional organizing efforts in the following years. A State Fair Grounds was developed on the east side of town, and in 1923 the Southwest Furniture Manufacturers Association organized the Retail Furniture Association of Texas, whose goal was to establish a market and rent temporary display space. Dallas did not have a monopoly of the furniture business; in the first years the markets rotated between Dallas, Houston, Fort Worth, and San Antonio. Given Dallas's ability to make the most of its assets, however, it might have had an edge on the other cities.

In 1936 some New York City gift manufacturers and import firms set up overnight showrooms in the Baker and Adolphus hotels in Dallas and then banded together twice a year to conduct a show. The gift industry is a nebulous concept which can encompass almost anything that isn't fixed to a building or nailed down, and even some of those items might be included. About the same time, several apparel manufacturers' representatives set up in Fort Worth. In 1937 they too moved to the Baker and Adolphus in Dallas to display their wares four times a year.

After World War II the Retail Furniture Association decided to expand into more ambitious undertakings. Two buildings on the Dallas State Fair Grounds were leased to house displays, and invitations were sent to buyers in Texas and nearby states. It was hardly in the same class as New York's or Chicago's, and it seemed likely to remain a modest operation, catering primarily to small manufacturers and merchants.

By 1948 there were several furniture factories in the Dallas-Fort Worth area. Both Kroehler and Simmons had established branch manufacturing plants, and there were several other small installations turning out lines of upholstered furniture. These plants were not rarities—there were nearly 150 factories in Dallas alone that year, turning out over $21 million in product.

For a number of years Dallas furniture manufacturers had discussed the possibility of creating either another mart by transforming an old building or by erecting a new one. Dallas seemed well suited for such a project because of its location. Atlanta, the regional center in those days, was too remote for the Southwestern trade, which was one of the fastest-growing in the nation. So the opportunity existed for aggressive Dallasites to win a new industry.

This was the situation when Crow became interested in trade mart ventures. Soon after remodeling the Doggett Building for the floorcovering companies, he conceived the idea of creating a more ambitious exhibition hall for Dallas. Why, he mused, should buyers for small department stores have to trek from one place to another to select their wares? In June, 1950, for example, the American Fashion Association, an apparel organization, six-day "Market Week" in which some 5,000 buyers from 16 states examined almost 900 lines of women's and children's apparel. Exhibitors took space in the Baker, Adolphus, Southland, and Bluebonnet hotels, and, in addition, local manufacturers displayed their wares in their own factory showrooms. In mid-week 3,200 people gathered in the Automobile Building at the Dallas Fair Grounds for cocktails, entertainment, and dinner. The spread-out display and meeting arrangements meant that buyers had to go from hotel to hotel, at a time when the temperature topped 90 degrees, try to find cabs to take them to the fairgrounds, and in general waste a good deal of time. Why not put all the display products under one roof, with one or two large hotels nearby to house visitors?

Crow reflected on the growth of Dallas and its development into the leading metropolitan hub in its part of the country. Business had never been better, and wholesalers were thronging to the city, which by then was becoming a major convention center as well. In time more exhibit space would be needed. Space for retail store merchandise would be a fine beginning, but there could be other exhibition halls later. Dallas was an emerging aerospace center; electronics were just starting to make an impact; and oil field equipment would soon be a mainstay of the city, along with apparel and insurance. Why couldn't there be exhibitions for at least some of these industries?

Moreover, it seemed clear that in time marts with permanent exhibits would replace annual or semiannual shows in many industries. Given high labor costs and other factors, it simply didn't make sense to bring in hundreds of tons of products to a single locale for just a few days. A large company wanting to make a splash might spend a quarter of a million dollars or more to set up an exhibit for a week or so. Why couldn't some enterprising developer rent space to that company at a few dollars a square foot for the entire year instead? And if others in the industry joined it, all could have a *permanent market* at a lower cost than the annual fair.

Crow's awareness of the possibilities did not come to him in a blinding vision, nor could he claim to see better than other developers what was happening in the city. Rather, there were a jumble of ideas running through his mind. By then Crow had become convinced that some new form of exhibition hall was needed, and that in time someone would create it. Perhaps that someone would be himself, perhaps not. The full scope of the exhibition hall possibility was not then apparent even to this optimistic developer. Initially he simply hoped to build a furniture mart and had some notion that other marts might follow. Again he operated on the principle of finding out what customers wanted and then providing them with a better version of it than they imagined possible.

In the middle 1950s, Crow began to think about what kind of edifice the furniture hall should be. Traditionally, a mart was a high-rise building, large enough to accommodate all who wanted to display their offerings and those who came to see and buy. The prototype was the Chicago Merchandise Mart, and while Dallas wasn't yet in that league or even considering such status, a smaller version might be ideal. Indeed, in November, 1948, the Dallas Manufacturers' and Wholesalers' Association considered plans for a 10-story mart that would have cost between $7 and $8 million. Plans were drawn up for what looked like a modest version of the Chicago Merchandise Mart.

Butler Brothers, which had been the preeminent merchandise wholesaler in Dallas at the turn of the century, owned a large building (around 670,000 square feet of space), down the street from the Doggett Building, that might have served Crow's purpose quite well. In 1951 Crow made another quick trip to Chicago, where he met with some of the leaders at the Merchandise Mart and received a short course in the operation. Returning home, he made an appointment to see the people at Butler Brothers to try to convince them to sell him their building so he could transform it into a mart.

Crow didn't get the deal. He lacked the experience and the kind of backing that the Butler people wanted. They turned to J. N. Fisher, who was thinking along the same lines as Crow and was better connected.

One can easily imagine Crow's disappointment at having lost this opportunity. For two years he had spared no effort in investigating the possibilities of planning, financing, and creating the mart, only to be rejected for what must have seemed unjustified reasons. In similar circumstances most businesspeople would probably have turned to other areas, possibly those in which success was more probable, to ease the pain and humiliation they had undergone. Had Crow taken that path, he might have wound up a major creator of warehouses in the Southwest—quite a distinction, but quite different from what actually happened.

Crow rejected failure then, as he would throughout his life. Crow often behaved as though he could will something to happen. On this occasion he managed to shrug off defeat. "It was probably just as well," he remarked, and then awaited the moment he would be able to realize his ambitions.

As it turned out, Crow's failure to win the bid for the Butler Building proved providential. Fisher started the renovation, which he projected would cost $1 million, in the summer of 1952. Everything that might go wrong did; there were cost overruns, and Fisher ran out of capital. He sold the building to Charles Sammons, a wealthy insurance man, and Sammons completed the job, turning the Butler Building into a place that displayed women's apparel, giftware, dinnerware, and a small amount of toys and housewares.

Thus Dallas obtained its own version of the Merchandise Mart in the center of the city, not far from the hotels where buyers could stay on their trips and where temporary exhibitors could put up displays. At one time such a building was exactly what Crow thought the city needed. But as he observed the opening of the new mart, Crow felt fortunate that he had not won the bidding. He became increasingly convinced that this downtown high-rise was the wrong building in the wrong location.

Although Crow had been impressed by Chicago's Merchandise Mart, he realized that it had been made for an age that had passed. Built by Marshall Field in 1934 at a cost of $28 million, the 24-story mart contains about 4 million square feet of space. Field planned to use it for his wholesale and manufacturing operations and to rent out the rest of the space, but it had been built at the wrong time—during the Depression—and lost money from the start. Joseph Kennedy purchased the mart for $18 million in 1946, and almost immediately it became highly profitable, since by then the postwar boom was beginning.

The venerable Merchandise Mart was (and to a degree is still) inconvenient, uncomfortable, and musty. The elevators of the time couldn't cope with the market traffic; customers ran up and down stairs to avoid long waits. And whether emerging from elevators or stairwells, would-be buyers found themselves in a maze of dreary hallways.

No one seemed to consider the inconveniences a cause for complaint. After all, the mart was clearly a place of business, not of pleasure. It was in a decaying part of the city, at a distance from the downtown area, inconvenient for those wanting to stay at the better hotels and enjoy Chicago's amenities. The Butler Building mart indeed was Dallas's version of the Merchandise Mart, and this was precisely what was wrong with it. Sammons had constructed and was operating a facility that was at least 20 years out of date. Drastic changes in the American business scene mandated a rethinking of old concepts. This rethinking is precisely what Crow did.

Crow would later remark that "the way of life deemed suitable today will be rejected as inadequate tomorrow," and that those who ignored this elementary fact were doomed to stagnation, if not outright failure. To his way of thinking, the Merchandise Mart was like the warehouses he had seen in Dallas when starting out in that field—undistinguished places designed for utility, without a thought to comfort or beauty. He asked himself, "Why shouldn't buyers be able to conduct business in a more pleasant atmosphere—open, airy, comfortable—and then at night return to a similarly constructed and operated hotel?" The age of Willie Loman, who tried to get by on a shoeshine and a smile, was over, but you couldn't tell it from the exhibition halls and commercial hotels in Chicago and Dallas. Such thoughts developed over the next few years, as Crow continued putting up warehouses, not only in Dallas, but in Atlanta and other cities as well.

In early 1952 Crow had a meeting with Bill Campbell, a real estate agent with whom he had done business. Campbell had been approached by 13 manufacturers and representatives of furniture companies who at the time were occupying old houses in the Cedar Springs–Maple Avenue area. They had correctly concluded that if they were all in one place, they could better serve Dallas customers and also draw others from nearby towns and cities. Campbell explained that the decorative furniture people had gotten the idea of a new, joint facility from a young man named George Hersman who had come to Dallas from Grand Rapids, Michigan, and at the time was selling a line of Widdicomb furniture from his house. Hersman wanted to rent display rooms, was dissatisfied with what he saw, and believed something better could be devised.

Hersman and Campbell collaborated in presenting the idea to several interested developers, and three—Charles Fritch, Irwin Grossman, and Crow—responded with bids. Crow proposed to erect the structure on Stemmons-owned land in the Trinity Industrial District. The request for bids was a chance to put some of his ideas into play, and Crow leaped at it. Together with John Stemmons he drew up a proposal to erect a different kind of design center.

Grossman was the lowest bidder and thought he would get the contract. He had made some errors in his computations, however, so a rebidding was called for. This time Crow won. The ensuing project, which he called "The Decorative Center," would mark a major new step in his career. Yet it would not have happened had Grossman been more careful in his estimates. (Crow rarely had to face such problems, for he took painstaking care in such matters as cost estimates. His ability to bring in projects on time and on budget—and at a profit—was becoming legendary.)

Crow's design center proposal was not for a single building, but rather a group of attractive, red brick structures with beige pilasters. The one-story buildings would be positioned around an off-street square or courtyard, pleasantly landscaped, with trees, benches, and other conveniences. Buyers could look out at the square as they went from one showroom to another, and instead of sitting on couches in dark halls, they could enjoy the outdoors when the weather was right, or take a break by walking across the square instead of through the buildings. There would be tree-shaded parking in the interior courtyard, and other touches to ensure that all who worked there were not only comfortable, but also could reflect on the superiority of the Decorative Center to other similar facilities.

The project would be impressive in physical as well as conceptual scope. The Decorative Center was to be larger than Campbell first suggested. The first two buildings would have 40,000 square feet of showplace space, but the third, designed by Anderson, would add another 60,000 square feet, enough to accommodate more than twice the number of dealers Campbell represented. The total cost was projected at $1.5 million, with occupancy set for the spring of 1955.

Negotiating the leases with the first 13 tenants was not as uncomplicated as it might seem. The lender required that each lease be subordinated to the loan, and Crow had to negotiate, and then renegotiate, with both lenders and tenants before finally signing them all to leases. Successfully leasing the third building involved a tremendous amount of effort. After having made many calls on the kinds of manufacturers deemed acceptable as potential Decorative Center tenants, Crow came to believe that a furniture mart building could be built for the manufacturers of volume furniture lines. In fact, he had to add to the center later on. Eventually it comprised six buildings, with 130,000 square feet of showroom space and parking for 200 cars.

The Decorative Center's proposed location at the intersection of Oak Lawn, Slocum Street, and HiLine Drive was not the ideal place for a showroom complex at the time, since access was somewhat difficult. The situation was changing, however. By 1955 Stemmons had concluded a deal whereby together with other landlords he would contribute 102 acres of land for an interstate freeway to be constructed only two blocks from the Decorative Center. Once completed, the freeway would facilitate rapid passage between the Trinity and downtown Dallas. Even before its completion, however, the highway prospects heightened interest in the property and enhanced its value.

The next step was to make certain the center was what he intended and promised it would be. Jake Anderson, who by now had his own company, was the architect. Anderson excelled in designing large but economical structures, which were what was needed since, as usual, Crow's finances were tight. Crow also utilized the services of landscape architects Arthur and Marie Berger, who planned a sunken garden between the buildings. Harold Berry, another local architect adept at keeping costs down, designed the subsequent phases of the center.

Crow had developed a strong interest in architecture and had definite ideas about how his buildings should look. After telling Anderson, Berry, and others of his desires, he would conclude with, "Take this idea and draw it up for me," or something like that. He instinctively mistrusted professional architects who operated on their own, without strong supervision from owner-employers. "Turn a building over to an architect and he'll ruin it for you," he asserted, implying that once he had learned the rudiments of the craft, he would never do so. Crow had also become aware of the tendency of some architects to prefer lavish treatments, since their fees were a percentage of the total cost. Some also view themselves as artists who have the good fortune of having client/patrons to pay any price for their visions, and Crow knew developers had to beware of this tendency as well.

The Decorative Center would prove to be a landmark structure, and it was an indication of Crow's developing taste. Then and later, Crow buildings were founded on his informed impulses. He was just a builder in this period, and he did not consider himself a city planner or architect, but his role and influence were expanding. A post-modernist without being aware he fit into any style "ism," Crow anticipated architectural trends that would surface elsewhere in the country two decades later.

While Anderson drew up the plans, Crow arranged for the financing. Together with the Stemmonses he organized the Center Realty Company, intending it to own and operate the center. Crow invested the grand sum of $950 for a 50-percent share, and the Stemmonses did the same for their 50 percent. The brothers provided the land in the Trinity Industrial District, which was sufficient equity to secure the mortgage, and Crow agreed to assume management of the project. Financing came from Mutual Benefit Life Insurance, and to cover a shortfall Eddie Kahn and John Stemmons added another $100,000 on an interest-free basis. Once again, the partners were operating with borrowed money. Their main investments were land, time, and reputation.

The Dallas Decorative Center, the first of its kind in the world to serve decorators and the design trade, opened formally on September 5, 1955. Compared to what would follow, it was quite small; on opening day it had 13 tenants. But to Crow and others it seemed magnificent. In writing of it that November, *Interiors* magazine said:

> *Young trees and clinging tendrils of ivy were softening the rose-beige paved walks, arcade soffit, and vertical Haydite louvers—as dignified as colonnades.*

Inside the showrooms, which were complete and functioning, cooled air and soft music cancelled the blazing Texas sun outside. The few easy steps from parking place to arcade, and the one-story layout which eliminated elevators and inevitably nerve-frazzling delays, were delights to the decorators and clients who have thronged the Center from 9 to 5 daily since September 6th, collecting samples from 25,000 square feet of showroom space in Building A, 13,000 in Building B.

Though the aesthetic virtues of the center were indeed appreciated by the tenants and customers, the magazine writer's implied "throngs" were an exaggeration. Even so, the Decorative Center was an instant hit. Crow opened a third building in 1956, the fourth in 1958, the fifth in 1965, and the sixth in 1967. In the spring of 1959, as old tenants expanded and new ones were added, the center was completely rented and had established a waiting list.

The Decorative Center and Campbell's input had stimulated Crow's creative juices. Even as the first buildings were going up, he was designing another, similar center for the furniture business—this one for the retail store trade rather than designers.

The buyers weren't particularly happy with conditions in Chicago. Hotels and restaurants were expensive, cabbies and bellmen were not all they might be, and the weather could be abominable. There seemed little they could do about the situation, however, and traditions die hard. Exhibitors and retailers had been going to Chicago for so long they tended to ignore the negatives, accepting them as the price of doing business. Any challenger would have to take into account the acceptance of the status quo. So the potential for another major furniture mart outside of Chicago existed, but the problems posed were formidable—indeed, far more difficult than Crow believed they would be when first broaching the idea.

Those with whom he spoke of a furniture mart rejected the concept out of hand. The Chicago and North Carolina interests made it clear they considered Crow an upstart. Retail merchants in the Southwest, and particularly in Dallas, feared that what would start out as a wholesale marketplace might develop into a retail discount center, creating serious competition for them. So the very retailers Crow hoped would become his customers were opposed to the notion. The furniture manufacturers might have gone along with the idea, since it would open other markets and put

Chicago on its toes. However, they were leery of the additional costs involved in leasing and staffing more major permanent showrooms, especially since, as far as they could see, Dallas would probably support a convention that lasted a week or so, but lacked sufficient business to justify year-round display.

The furniture people did not scorn Dallas; in fact, most of them showed up at the annual furniture show put on in the Automobile Building at the Fair Grounds. For this once-a-year event they packed their furniture onto railroad cars, shipped the loads to Dallas, and then, after the show ended, either sold the display wares at deep discounts or took them back to the factory. The logistics were almost as complicated and frustrating as those the clothing buyers and exhibitors had to manage in 1950. Crow calculated that they could have a permanent exhibit at what was to become the Homefurnishings Mart for around the same cost it took to mount their exhibits at the Automobile Building. Besides, the new building would be more accessible, more attractive, and air conditioned, unlike both the Merchandise Mart and the Automobile Building.

The structure Crow had in mind would be larger than the furniture interests would require, for he felt reasonably certain that his carpet tenants in the Doggett Building would leap at the chance to come in on the deal. His strategy was to reach a critical mass as soon as possible. Once it became evident that the Homefurnishings Mart would be attractive due to its clientele, additional manufacturers would want to participate, which would make it all the more appealing, and then the cycle would continue. It was obvious to Crow that the Chicago Merchandise Mart had enjoyed success not only because it was a great edifice, but also because its tenants knew they had to take space since everyone else was there. That "must" status was what Crow aimed for at the Homefurnishings Mart. And that was why it was planned as a large structure, 434,000 square feet, to be constructed in two stages, the first accounting for 217,000 square feet. At $3.5 million it would be by far the most expensive project he had undertaken.

John Stemmons was pleased to come in on a deal that clearly would benefit the Trinity Industrial District. In return for providing a million square feet of land, the Stemmonses would receive a half interest in the structure. Equitable Insurance agreed to make the loan on condition the building be 70 percent rented with half of the leases for five years or more, at a rate of $2.25 per square foot. The Mercantile National Bank in Dallas and the First National of Boston agreed to finance the Equitable takeout. So the financing and the land were in place. Everything was contingent on getting signed leases.

Crow took to the road trying to win over the furniture interests. Asking an entire industry to shift its focus to a new locale was an audacious undertaking. The going was tough. The challenge was to convince the furniture manufacturers of the United States, but especially North Carolina, that a Texas businessman unknown to them, with no experience in furniture, had the ability to realize a vision, and that they should gamble that he knew what he was talking about. Crow persevered, not just as a matter of blunt determination but with confidence that his idea was right.

At times winning tenants for the proposed mart seemed impossible. Crow and Bill Campbell traveled to North Carolina, living frugally in double hotel rooms and avoiding restaurants because they didn't want to spend scarce capital. They learned what rejection meant from dozens of polite but firm Carolinians, and some who weren't so polite. "I hope you don't build it," one of them told Campbell, "but if you do, we'll do everything we can to ensure it won't be a success."

At the end of one particularly discouraging day of going from office to office with nothing to show for it, the two men sat on their beds, looked at one another, and wondered what would happen next. "What do you think we ought to do after this trip?" Crow asked, and Campbell, utterly dejected, replied, "We're licked. We might as well admit it, go home, and not waste any more money." Crow knew they hadn't accomplished anything, but the idea of being defeated never occurred to him. The mart concept seemed so obviously beneficial to everyone concerned that he was sure that in time they would strike fire. Crow gazed searchingly at Campbell for a moment, and then stood up and wagged a finger in his associate's face. "I'll tell you what we're going to do," he declared. "We're going to build the finest furniture mart in America. I'm going to build it and you're going to lease it." Then, as though to underline his confidence, he lay down in bed, turned on his side, and was asleep in a matter of minutes.

They returned home without leases, but with a determination to try again. At least some of the people they met had said they would withhold their final decision until they saw the building, and Crow hoped others would feel the same way. He believed that once manufacturers went through the novel mart he was planning, most would sign leases.

For the next few months Crow devoted his time to the mart's financing and design. The former was a constant struggle, the latter much more enjoyable. The Equitable would not fulfill its commitment unless he had signed leases. Crow turned to the First National for interim financing, but they would give him nothing until the Equitable's requirements were met. It was a Catch-22 situation. No money, no building. No building, no leases. No leases, no money. And then round again.

Crow tried to get some support from the Republic Bank, where John Stemmons was a director, but he was turned down by the loan committee, which also wanted signed leases. He then went to his old associate at the Mercantile, J. D. Francis, whom he had known since the days when Crow was a runner and teller there. Francis was on his way to a convention in Chicago, but he gave Crow a hearing and said he would see what could be done. At the convention Francis arranged a meeting with Bill Keesler of First National of Boston, and told him he intended to participate. Keesler agreed to go ahead with the commitment. So the work could proceed.

Crow and Harold Berry, the Furniture Mart architect, planned a two-story structure, with such humanizing touches as deep bays for lounges, broad corridors, and large apertures between floors. It was to be constructed in two phases, each with 217,000 square feet, and the total cost was estimated at slightly more than $3.5 million.

With the Homefurnishings Mart well underway, Campbell and Crow once again visited the manufacturing centers, especially those in North Carolina, seeking additional leases. Crow went so far as to hire a public relations specialist, the wife of one of the furniture representatives. He thought she could build ties with the retail furniture industry. But she had no confidence at all in his ability to succeed, and told him so on several occasions. The retailers would fight him tooth and nail, she advised. The manufacturers would not come because of fear of losing business. There was just no way he could prevail. After a while Crow was fed up with this "professional" advice. The P.R. expert was dismissed, and he wouldn't use another one for many years.

Crow persisted in seeking tenants as the Homefurnishings Mart continued to grow, and he finally got more of those leases. Ground had been broken in 1956. While the mart was being built, Margaret and he would take their children for Sunday picnics to the construction site. Trammell would sit there, stare, and grin. The new mart was turning out fine. The world's most ambitious display complex, the Dallas Market Center, was well under way.

The problem of inaccessibility, which was supposed to be solved by the bridge, was resolved by another development and became moot. Construction began on Route I-35, the first project completed under the terms of the Federal Highway Act. A service road to the Homefurnishings Mart was given priority and was the first section to be completed. The entire expressway was scheduled to open in the late 1950s, and in fact its grand opening took place in December, 1959.

The freeway stimulated development efforts in the Trinity Industrial District, and so was of enormous importance to Crow and the Stemmonses. With the freeway in place, the mart would be less than 10 minutes from downtown Dallas and only 15 minutes from Love Field. "It was the best business deal I ever made," John Stemmons often characterized his donation of the land. "I knew that construction of the freeway would benefit me and the city because it got people from here to there." From then on the district underwent a boom in which Crow and the Stemmonses participated.

Crow would say that nothing he had done gave him such a feeling of professional accomplishment and sense of success as did that mart. He gave much of the credit to the Stemmonses, without whom Dallas would not have had the freeway. He proposed that the highway should be named after the brothers, but traditionally such project names honor dead people. So Crow suggested the road be named after their father, Leslie, and so it was.

A characteristic incident generated positive publicity for the Homefurnishings Mart, leading to increased business. One of the first tenants informed the building manager that the wall being erected between his space and his next-door neighbor's was incorrectly located. The manager told him the error would be rectified, and called the contractor to tell him of the problem. Such responses could be expected to satisfy the tenant. Crow happened into the building soon after and ran into the tenant, who repeated the complaint to him. Together they walked to the space, and Crow was shown what was wrong. Having made sure the complaint was valid, he asked some workers who were on the site to take down the offending partition. When they seemed reluctant to do so right away, Crow grabbed a large sledgehammer and began the demolition on his own. The workers got the idea, and within a few minutes the wall was down.

This dramatic reaction may seem a bit foolish. After all, the tenant was content to know the wall would be down in a few days. But now he could go home knowing the job had been accomplished. He also knew that Crow had demonstrated, quite convincingly, the extent of his concern for his tenant's well-being and peace of mind. The story got around Dallas the following day, and in the weeks that followed the rate of rentals increased significantly.

The first half of the mart was inaugurated, fully leased, in July, 1957, and the doors opened to 1,850 buyers the first day. It was terribly hot. When the exhibitors turned on all their lights, the transformers blew, killing the air conditioning system. There wasn't a window in the place, and everyone was sweltering in the humid heat. Crow quickly had the transformers repaired and then packed in dry ice to keep them as cool as possible. Workers manned the transformers that day and into the next.

Even as the first customers streamed into the initial section of the mart, work began on the second section. By the time it was completed in January, 1958, some 90 percent of its space had been leased, and there was a waiting list of manufacturers.

Such "continuous building" was the norm in several of Crow's other marts. He began by erecting a structure about half as big as he thought it eventually would have to be, started leasing even before construction began, and then, when demand warranted, commenced work on the second part as soon as possible. Those who considered Crow overly optimistic and a risk-taker did not reflect upon this strategy, which gave him a conservative fall-back position.

The situation seemed to support careful optimism. Crow had become convinced that there would be a demand for several marts at the Trinity, that each would augment and support the others, and that a hotel—perhaps two or three—would be needed to house customers and exhibitors. The obvious success of the Homefurnishings Mart supported this conviction. Perhaps Crow was giddy with the accomplishment, but even as the second phase was beginning, he regretted not having made it larger and vowed not to make the same mistake with subsequent marts. The next time out the first section would be larger.

The "next time out" was The Trade Mart, a four-story, $7.5 million structure on which work began in 1958. Like the Homefurnishings Mart, it would be constructed in two stages. It would accommodate additional furniture showrooms and in addition would be suited to the gift industry. The Trade Mart opened on February 22 of the following year.

Crow intended The Trade Mart to entice manufacturers and buyers of housewares, electric fixtures, and the like. Dallas already had some showrooms for this industry, primarily in the old Butler Building that Sammons had renovated. There was an exhibition space in the Santa Fe Building as well, which had been renovated in 1947 in a bid to become a new wholesale giftware mart. Finally, once a year a gift show was staged at the Baker Hotel. What John Stemmons and Crow had in mind was much bigger than all three locations combined. Their Trade Mart would have a million square feet of exhibition space.

A dramatic feature of The Trade Mart's space was an atrium, the first in that kind of building in the nation. It was a device Crow was to use many times, most dramatically in his hotels. Atria are named after the Roman town of Atria, where they are said to have originated more than 2,000 years ago. They were interior courtyards in the homes of wealthy citizens who centered their family lives around the private garden areas. Atria continued to be popular in Mediterranean lands (Crow was influenced by one he saw in Italy), and were brought to the United States by the Spaniards and the French. They were particularly popular in the French Quarter in New Orleans, for example, where the drab and windowless outside walls belie the green serenity within.

The inward-looking characteristic of the atrium, Crow thought, might be well suited to public buildings as well as to private homes. People in conventional public buildings want windows to look outward—away from the business at hand. A corner office in a commercial building is desirable because its occupant can see more of the world through the two windows. In an atrium-centered building, however, that same person might want windows facing the park-like expanse of open space. Psychologically he or she would be oriented *into* the building rather than out of it, an orientation that was particularly important for a mart, where all activities are directly or indirectly intertwined.

Crow never claimed to have invented the atrium, of course, and incorporating one into a public building was not new to America. The Brown Palace in Denver, constructed in 1892 and still operating, has a small nine-story atrium with rooms arranged around open corridors overlooking the lobby. The Palace in San Francisco, constructed even earlier, also had an atrium, but it had been demolished in the earthquake and fire of 1906. In the late 1940s, Crow was to learn, a West Texas oil man asked Frank Lloyd Wright to draw up plans for an office building with an atrium.

The atrium concept was a good fit with Crow's visceral dislike of the closed nature of most showrooms and marts and his desire to make them more open, accessible, and airy. An atrium would give his million-square-foot, windowless building a humanizing element. "The entire space was conceived to make the buyers' trips to markets more enjoyable and restful," said Crow.

When Trammell and Margaret visited Milan, Italy, they had seen a great old public building with a doughnut-shaped courtyard in the middle, and Crow was struck with how simple it would be to span that courtyard with a sloped roof and make it into an inside space. He later decided his atrium in The Trade Mart would be about 110 by 450 feet, extending all the way to the top of the structure, with a glass cover. It would be simple enough to accomplish technically. The building was to be a large square. So essentially it could be cut in two, the halves moved 110 feet apart, sections added on either side to contain the hollow square, and all enclosed with a roof. After that, simply paving the inside floor produced the atrium. The space would be lighted by the sun in the daytime and heated by the exhaust from the showrooms.

The Trade Mart atrium would not be just empty space. Crow included many useful features—balconies, crosswalks, and escalators at either end, for example. Building the atrium added a rather modest $750,000 to the cost of construction. Crow and Stemmons passed along the expense to the renters, who were willing to pay more than the usual rate once they appreciated the unique benefits the atrium contributed to their place of business.

The atrium was not The Trade Mart's only unusual element. For instance, the mart was the first of its kind to devote space to both permanent showrooms and temporary booths. The plan was to have some 600 permanent exhibitors in the huge building, depending upon space requirements. The entire fourth floor was to be devoted to furniture, thus relieving pressure on the Homefurnishings Mart and Decorative Center, both of which had waiting lists.

Getting companies to take space in The Trade Mart presented different problems than those encountered with the Homefurnishings Mart. This time Crow felt there would be no problem convincing industry leaders he could fulfill his promises—he had done so already; and there was no single, coherent industry involved. Rather, the initial task would be to win renters from the other, older marts in Dallas, and then go on from there to attract outsiders.

It soon became evident that Crow had underestimated the difficulties with leasing. Perhaps he was too optimistic; the overflow from the Homefurnishings Mart didn't go far toward filling the space. Then the warring factions in the Sammons operation and the Santa Fe got together, with some of the Santa Fe people moving into the Dallas Merchandise Mart instead of into his building. A massive leasing effort had to be mounted, and to head it up Crow again brought in Bill Cooper, who was one of the best salesmen he knew. They hired Horace Ainsworth to join Cooper. Grady Jordan and Gibby Ledyard from the J. W. Lindsley Company were added. Jordan concentrated on the housewares, toy, and floor covering people, Campbell on furniture, and the rest covered the field. It was the largest sales effort Crow had yet undertaken.

The pressure was on, for the sales force knew they had to sign sufficient leases each day to bring in enough revenue to service the debt. Crow had to develop and then implement a strategy to win clients in the face of opposition from some of the most powerful and entrenched interests in Dallas. He was also seeking to shift the center of gravity for one of the nation's larger industries. The challenge called for resourcefulness, imagination, tenacity, and a dollop of good fortune.

The first step was to go after the local Dallas market, which meant winning tenants then in the Santa Fe Building and the Merchandise Mart. Members of the sales team visited lessees personally, to woo and, they hoped, win them to The Trade Mart. There were complications. When management of the Santa Fe learned what Crow was up to, it ordered the agents off the premises. They tried again and once more were ejected. Crow tangled with the officials there. "This is a public building," he argued. "There's a public business being conducted here, and we're here doing business. We acknowledge your concern, but we *are* going to solicit exhibitors from this building." The Santa Fe management disagreed; again they were thrown out. When Crow did manage to see displayers, some showed anger at his audacity in attempting to disrupt their long-standing arrangements. One shouted that he would sooner go out of business than take a lease with Crow, going on to say that he would organize others who felt the same way.

They met similar opposition at the Merchandise Mart, where the manager of the gift show ordered two uniformed policemen to evict them. Crow complained that this wasn't lawful; they were ousted nonetheless. Perhaps he should have taken legal action, but he felt the circumstances didn't call for it. Crow reflected that a businessman is in trouble anytime he worries about a competitor talking to his clients, and he is in good shape when he knows that his competitors' clients will seek him out for a better deal. Crow felt he was in the latter position as The Trade Mart neared completion. He believed he would get the tenants, for they were learning that he was going to have the best building by far, and at competitive rents. Besides, he already had the furniture industry, and furniture stores are significant purveyors of gift and accessory lines.

Crow, Campbell, Cooper, and Ledyard also went to Chicago to seek out clients. Several shows were in progress. The four stayed at a suite in the Sheraton Towers in order to have an impressive place to meet potential customers. The bellboys thought their behavior was strange, but the men stayed there the length of the shows and wrote leases every day.

Financing remained a constant problem. As before, the Stemmonses and Crow put little money into the project, hoping to rely on borrowed funds. The Equitable handled the mortgage, but financial pressures continued. The Furniture Mart was throwing off some earnings, but hardly enough to cover cash flow problems at The Trade Mart, which was still rising out of the ground and had a million-dollar deficit before it was finished. Crow obtained a $300,000 loan from the James Stewart Company, his general contractor on the job, and $100,000 from Harmon Electric and $50,000 from Beard Plumbing, who were in on the construction. Dallas Power & Light helped by purchasing the electrical distribution systems in the building and metering each tenant directly. Crow also asked General Electric to "lend" him several hundred dollars' worth of air conditioning equipment until he could get the mart going. The plan was approved by General Electric Credit Corporation, but only after G.E. itself guaranteed repayment of the debt. In addition, the head of G.E.'s local dealership, Texas Distributors, had to personally pledge his business against a possible default. Finally, Margaret and Trammell signed a $300,000 personal note based on her holdings, meaning they were risking Margaret's inheritance on the project.

Crow opened The Trade Mart in January, 1960. It was an auspicious occasion, for they dedicated not only The Trade Mart, but the Stemmons

Freeway as well. Construction was still underway when the ceremonies began. Mayor Thornton's speech was fitting—but when he stepped from the platform he sank up to his ankles in wet sand.

Although Berry designed The Trade Mart, Crow had brought in another architect, Harwell Hamilton Harris, to help design the atrium. Harris devised plans for what they called the Grand Courtyard. At one end there was to be a redwood gazebo overlooking a rustic pool, and at the other, a huge fountain that thrusts its sprays higher than the fourth-floor balcony. There would be palm trees, shrubs, vines, flowers, and colorful flocks of parakeets and finches flying around the upper floors.

Problems appeared immediately. Crow insisted on having real birds flying freely, and he purchased every parakeet and finch for sale in the city. They were turned loose, and off they flew down the corridors, into the showrooms, all over the building, leaving their droppings wherever they went in what the local wags were soon calling "the largest birdcage in the world." The birds were gone in a week, and Crow tried again. This time the birds were kept caged for a while before being turned loose, and they stayed where they were supposed to.

That wasn't the end of it. Not content with their birdseed diets, the parakeets and finches started to eat the plants. Also, bird lovers volunteered to supply additional varieties, and in came ducks and doves. They were beautiful but troublesome. For example, there was the time a white dove unloaded on Storey Stemmons while he was delivering a speech, and the time a goose started to gnaw at the leg of Mayor Thornton during a dinner. The doves went.

The bird flutter not only led to free publicity; the winged novelties were part of a milieu that made The Trade Mart an interesting location in which to conduct business. (Crow liked to think it was done that way in the Rialto five centuries earlier.) A sidewalk restaurant around the fountain offered visitors a comfortable, peaceful location in which to eat their lunches. Today the atrium remains an altogether graceful element providing exhibitors and customers with a place where they can rest, converse, and revive themselves, or where customers can simply sit and relax—after which they can go back in and buy.

Tenants and potential tenants knew that Crow and his associates were striving to develop ways to serve them better. One innovation was the institution of boards of governors for each industry. The boards would provide direction in market policies; in effect, the renters would have a forum at which they could state their needs. This participation both ensured the appropriateness of Crow's policies and gave confidence to the industries that matters were being handled in their best interests. All marts have such boards today; Crow was the initiator of this practice.

The Trade Mart's success was due in large part to the eventual willingness of tenants at the Santa Fe and Merchandise Mart to relocate. When they saw the benefits of operating from the new facility, many abandoned their earlier opposition and embraced the new structure and method of conducting business. After a while the older buildings were practically deserted. One of the last holdouts was the tenant who earlier had sworn to oppose Crow to the very end. One day Crow entered the Santa Fe and knocked on his door. There, behind the desk was the dour businessman. He looked at Crow, scowled, and asked, "Where's the lease?"

Crow's first two marts gave him much pleasure, not only because they were profitable, but even more as the realization of his vision. Throughout this period he continued to erect warehouses and other structures, but the feeling of accomplishment these provided was nothing compared to that of the marts. Indeed, by the mid-1950s Crow was putting up warehouses in places he hadn't seen and would never see. But the marts changed the face of Dallas, and for this reason Crow was becoming known throughout the nation.

The marts enhanced Crow's credibility as a builder and developer. Whenever anyone doubted his ability to carry through on promises, Crow would refer them to the marts. In time it wasn't even necessary to mention them, since all who dealt with him knew of these and later accomplishments. The credibility factor was important, because real estate developers require credit in order to function, and Crow's ambitions were such that he never had enough money available for all his projects. Furthermore, as anticipated, the marts had an initial negative cash flow, and until this was turned around he had to tread carefully.

By the early 1960s Crow had established a solid position with the gift industry and was having an easier time obtaining renters in that business, but otherwise he still had to scramble for clients. Cooper had traveled to New York, Boston, Los Angeles, San Francisco, New Orleans, and smaller cities and towns, seeking out gift tenants. Crow had also hired an advertising man, Hal Copeland, to locate temporary renters who would take space only during the three gift shows each year. Despite these efforts, it wasn't until June of 1959 that The Trade Mart reported a positive, though small, cash flow, and by 1962 the marts were liquid for the first time.

The following year The Trade Mart became associated with tragedy. President John F. Kennedy was in Dallas on November 22, 1963, to deliver a speech at a luncheon at The Trade Mart. Three thousand people were in the atrium awaiting the arrival of the Kennedys, the Johnsons, and the then-Texas Governor John Connally. Crow recalls what happened:

I had been given the honor of greeting the President and his party as they arrived in their cars, and to escort President Kennedy to his place on the platform. Standing there alone, by the curb, having been advised by the Secret Service that the President's party would arrive in minutes, I was surprised by quick, sharp taps on the shoulder. It was a Secret Service officer, who said, "Something has happened, trouble, but we don't know just what. Stay here! We don't want the crowd to react! The caravan will drive on by. Please be steady and calm."

In less than two minutes the caravan did appear, and did drive on by, and at a fast speed. Being only one lane from me, I well saw the President, leaning over onto Jackie Kennedy. The party was distraught, obviously, but I do not remember seeing blood. I went back into the hall. A pall of agony had settled over the room, and people were openly weeping. I made an announcement that the President had been shot and his condition was unknown. Within the hour we got the news that the President was dead. In anguish and despair, people left the hall.

The next building in Crow's ambitious complex was called the Market Hall, which was to be erected across from the Trade and Homefurnishings marts on Industrial Boulevard. The genesis of the Market Hall came in 1962, when it appeared the National Lumbermen's Association might relocate its meetings, scheduled for November, 1963, due to lack of space. Meeting with Bill Cooper and the Stemmonses, Crow decided to go ahead with his next project in the Trinity Industrial District. Cooper put together figures that demonstrated the need for a 114,000-square-foot, single-story structure and another 44,000-square-foot, two-story building. The only problem was that there might not be enough business to run at a profit. However, the Stemmonses and Crow calculated they could use a large bookkeeping loss at that time for tax purposes, and so they went ahead with the planning, out of which came the $2.5-million building, which today contains 214,000 square feet. The Market Hall is the largest privately owned exhibition hall in the United States.

Market Hall was designed for use by temporary exhibitors, conventions, and the like. Creating a structure for this purpose made abundant sense. For example, some furniture manufacturers had no desire or perceived need for permanent space in the mart but would have liked to be represented at the regular conventions held by the industry, and the same was true for companies in the gift field. The Market Hall was Dallas's attempt to capture some of the business that went to Chicago, New York, and elsewhere.

Crow's idea was to bring temporary exhibitors from throughout the country together with the permanent showrooms at the Market Center. The earlier buildings would draw permanent exhibitors to the center, which in turn would be enhanced by the Market Hall. In this way he could strengthen both elements and the markets overall, and in addition give himself an advantage over those other cities. He called the transient exhibits the frosting on the cake. The mix worked, and mart operators in other cities imitated him where they could.

Crow was in competition with the municipally funded and operated exhibit center, a part of the Dallas Convention Center, which had only about 150,000 square feet of exhibition space at the time, but which was expanded to 400,000 in the mid-1960s and to 600,000 soon after. Even so, Crow won business from the municipal operation because his concept of a marketing complex worked so well.

Stemmons had always insisted that each building stand on its own financially, and therefore the various holdings were incorporated separately. The Dallas Market Center Company had been incorporated in the early 1960s to serve as a management umbrella for the existing markets. It was merged with the Homefurnishings Mart, The Trade Mart, and Market Hall in 1963, and at that time the four corporations became united.

Everything was proceeding so well that Crow was spinning off ideas for new marts all the time. The next was one for the apparel industry, whose local business at the time was centered in the Sammons Merchandise Mart. Breaking this near monopoly presented Crow with legal problems, the first important ones he had to overcome, providing him with a lesson on how to get things done in the face of legal roadblocks.

Joe Ragland, general manager for the Sammons Mart, seemed to have Dallas's wholesale apparel industry in his control through a long-standing agreement with an organization of apparel salespeople known as the National Fashion Exhibitors Association (NFEA). This agreement provided that Sammons would rent apparel exhibit space to none but National Fashion members, and, in turn, these members would not exhibit anywhere else in Dallas. It was a protected arrangement. No competition was envisaged, and none would be tolerated. This *de facto* monopoly effectively functioned as a restraint on anyone who might be able to offer a better product at a competitive price. Namely, Trammell Crow.

Yet there was a flaw in the monopoly. Another organization, the American Fashion Association (AFA), whose members were drawn from other apparel manufacturing companies, exhibited in the hotels. There was considerable tension between the sales representatives in Sammons's mart and the manufacturers they represented who were exhibiting in the hotels. Sammons refused to rent space in the Merchandising Mart to manufacturers. This meant that if a line was dumped by a sales representative, the manufacturer had nowhere to go except to the hotels, where he would have to take the least desirable location until, over the years, he developed some seniority in the AFA. The manufacturers–salespeople rift was a small opening through which Crow might work, an angle to play.

Some of the salespeople who leased space in the Sammons Merchandise Mart felt that the exclusion of manufacturers was detrimental to the growth of the industry. Moreover, they realized Crow might be able to break the strangleholds of the associations. Understanding what was happening, one of John Stemmons's lawyers warned Crow that he might expect a lawsuit should the tenants break leases and come over to his new complex. Additionally, there could be tenants' suits claiming the partners' actions had made their leases less valuable. The amounts demanded could be in the millions, the attorney warned, and should be considered very carefully before acting.

Crow and Stemmons consulted Henri Bromberg, who was Crow's lawyer as well as an investor, and were dismayed to learn that he agreed with the legal disaster scenario. "I've studied this carefully," Bromberg said, "and in order to do what Trammell wants, you're going to have to go in and get all these tenants in the Merchandise Mart and the Santa Fe Building. When you start pirating all those tenants, I think you're liable to be sued for interference with contracts, and that you'd be in significant jeopardy, both as to risk and the outcome and the amount."

Crow didn't know how to respond. John Stemmons turned to him and said jokingly, "Well, Crow, this ends our little fiasco. We're through Crow. We ain't going to have any part of it. We've enough trouble without getting into a lawsuit." Stemmons later reported that he and his brother had to lead Crow out of the room. Crow was in a half-daze, so great was the blow. But he didn't give up. Crow believed that when in doubt, you should seek another opinion—and keep going until you find a sensible one that agrees with your convictions.

So Crow went to see another of his attorneys, Eugene Locke, partner in the Dallas law firm of Locke, Purnell, Boren, Laney and Neely. Locke and Joe Stalcup, a young attorney with the firm, agreed that Crow might lose the court battles, but Stalcup thought the liability couldn't exceed $1 million, and even with losing the case, with some luck, $250,000 would serve. Crow returned to the Stemmonses and asked if they would accept Locke's opinion. John replied, "Hell no, it won't suffice for me. When Dick Scurry [another lawyer] tells me by God that that's the way it is, then I'll go with you, and until then forget it." Crow asked Locke to talk about the matter with Scurry, and after a while Scurry came round and told the Stemmonses of his agreement with the Locke firm. As John Stemmons recalled, "There was great rejoicing in Crowtown, and away he went to build the Apparel Mart." By the time construction began in mid-August, 1963, Crow had signed 193 tenants from competing buildings.

The Apparel Mart was constructed north of the Homefurnishings and Trade marts. Knowing that the apparel industry is quite unlike the furniture and gifts businesses, Crow set out to build a structure that would be at the same time familiar and different. This was not the kind of job for Berry, who was terrific when it came to keeping costs in line but did not excel in designing glamorous buildings. Crow issued an invitation to submit concepts, and five firms sent in plans. All were told what was needed. "We've got to have a courtyard that will be Miami Beach modern," Crow told them. "This is the apparel industry and the interior has to look like the apparel industry."

And so it did. The competition was won by Pratt, Box, and Henderson, a Dallas-based firm that had some experience along the lines Crow required and was young and hungry enough (average age: 33) to give him what he wanted. The atrium of the Apparel Mart, known as the Great Hall, is flanked by balconies overlooking it from four floors. Crow installed a large stage with permanent lighting and sound equipment to create spectacular settings for fashion shows, and on the third floor there was a smaller theater for more modest productions. Then, to make certain the place had the class and cachet to win customers, he hired Kim Dawson, who was the fashion director for the Sammons Merchandise Mart and who, Crow had been told, was the most qualified person around.

Meanwhile, the anticipated lawsuits were never filed. Stalcup's prediction proved remarkably prescient. Claims arising out of the transfer of renters to the Apparel Mart were settled for around $250,000. Crow was most grateful for his assistance, and the new firm that Stalcup founded did most of Crow's legal work from then on. In fact, that episode was to alter the course of Crow's life, for one of the young attorneys in the Stalcup firm was Don Williams, who succeeded Crow as CEO some years later.

The Apparel Mart opened in October, 1964, with 880,000 square feet on a 20-acre site. Several expansions followed, and by the early 1980s it had a total of 1.8 million square feet, larger than 37 football fields, and was six stories high. It is the largest building of its kind in the world.

As had become Crow's practice, he was busily planning the next mart even before ground was broken for the Apparel Mart. The thought of a new undertaking was almost too much for the Stemmonses, who not only had been shaken by the Apparel Mart experience, but were having qualms about Crow's way of conducting business. John Stemmons confided that his brother Storey "always figured we had to keep ourselves pretty liquid because this wild man is going to go broke." he added, "All of a sudden he's going to make a big pull on us for a lot of money, and we're going to have to be able to reach in and get it to keep him from going under. So Storey got really concerned."

In 1959 Storey Stemmons arranged to have Bill Cooper (who by then was general manager of the marts) assigned to membership on the World Trade Committee of the Dallas Chamber of Commerce. At one session a member suggested that Dallas's international trade position would be greatly enhanced by a building to house importers, exporters, consuls, and trade offices. Another suggested this would fit in well with "what Crow is doing in the Industrial District." Cooper presented the idea to Stemmons and Crow, they agreed, and in 1960 the World Trade Center Company was incorporated—but not activated.

While prepared to take the initial step, Crow and Stemmons were reluctant to proceed beyond incorporation until they felt reasonably certain they could get tenants for the new building. Both men suggested to Cooper that in his spare time he try to firm up letters of intent from prospective tenants, which he attempted to do. The project was difficult to sell. As Cooper was aware, embassies throughout the world had been approached with blueprints for buildings that were never realized. Those that had convinced their governments to take space in ephemeral buildings had become gun-shy and insisted that the space be in place before signing leases. For this reason Cooper suggested to Crow that the international mart be erected between the Homefurnishings and Trade marts, utilizing the upper floors as additional mart space and devoting the ground floor to the World Trade Center. In effect, the old businesses would subsidize the new.

Nothing happened for a while, as Crow concentrated on other matters. Meanwhile the tenant lists grew at the other marts, and it soon became evident that unless he found some way to accommodate those firms waiting to come on board, they might be tempted to go elsewhere.

From 1968 to 1971, Cooper dutifully related in his annual reports that the international project was due to start soon. In the summer of 1971, prospective tenants were beating on his desk, saying they were desperate for space. Rumors abounded that other developers were giving the matter their attention, and Cooper worked up an appealing *pro forma* plan for the World Trade Center. In July, 1972, Crow selected Beran & Shelmire, a Dallas architectural firm, to design the center. It was to be a 1.5-million-square-foot building.

Crow approved of the proposal, and he and Cooper met with John Stemmons to urge him to join in the deal. Storey, who had been the more venturesome of the brothers, had died the previous year, and thereafter John had become more conservative. He now told Crow and Cooper he was willing to go along with a $25-million project assuming a loan at 6-percent interest could be had. Stemmons knew this financing was out of the question; rates at the time were running around 9 or 9½ percent. High double-digit rates lay in the future, but neither Crow or Stemmons could have predicted them. At the time, anything over 6 percent seemed usurious to Stemmons, who in any case was becoming increasingly nervous about Crow's actions.

John Stemmons is a salty and direct man, and one observer of the meeting recalled that John looked rather angry and shouted, "Crow, I'm not going to have any part of any such unconscionable money-lending son-of-a-bitches that would try to charge me 9¼ percent. I'm not going to do it." Crow said he would see if he could do any better, knowing that it wasn't possible.

Crow was sure an international center would inevitably be built; if he didn't build it, someone else would, perhaps to the detriment of his and Stemmons's mutual interests. One Sunday he invited his long-time associate to his home to discuss the matter. Stemmons agreed with Crow's reasoning, but simply couldn't accept paying 9¼ percent interest on so large a sum and having to meet such heavy mortgage payouts. Six percent was tops, he said, inveighing against "those god-darned moneylenders in New York who bleed you for every dime." The two talked of getting around the problem by reducing their interest in the World Trade Center from a half share each to a third and selling the other third to outside investors, with the extra cash being used to erect the building. They also considered taking the entire Market Center public and putting up the new building with part of the proceeds. These alternatives were not appealing.

Clearly, as both Crow and Stemmons recognized, they had reached an impasse. Neither had the heart to back down. Stemmons simply looked at his partner for a moment that Sunday and said, "Crow, I'm going to go back and figure what our part of this deal is worth, and then you can buy us out."

Crow agreed, and each man put his accountants and associates to work on evaluating the Market Center partnership prior to its dissolution. Working for Crow, Cooper came up with the figure of $16 million. George Shafer, the executive vice-president for Industrial Properties, thought it should be $14 million. Using Shafer's calculations, Stemmons agreed to sell for $7 million, but Crow, with Cooper's estimate, was prepared to pay $8 million.

What happened next is a good illustration of the way Crow and Stemmons conducted business in those days, reminiscent of Crow's attempt to assume the losses on the earlier deal with Kahn and Bromberg. Stemmons said, "Shafer says our equity is worth $7 million and that's what you're going to have to pay for our half." Crow countered, "Well, that's not enough." Stemmons would have none of this. "Now, Crow, this has been a hard goddamned thing for me to do, to sit here and break up this partnership, which has been so wonderful and I've enjoyed it so much. And we ain't going to argue price. I'm giving you the damn price, and all you can do is say yes or no."

So began a bizarre tug of war. Crow insisted on paying $8 million and Stemmons insisted on receiving $7 million. At one point Crow told Cooper that when it came time to write the check it would be for $8 million, and that would be that. Stemmons called Cooper in and told him that if Crow tried to give him a penny more than $7 million he would tear up the check and the deal would be off. Crow finally acceded, and the deal was consummated on December 19, 1972. The Market Center thus became a Crow family affair, with the land remaining the holding of Industrial Properties, leased to Crow for 99 years.

Construction soon began on the World Trade Center, which provided 1.5 million square feet of space and cost $75 million. As will be discussed later, there was more to come at the Market Center, which has completely shifted the city's focus. But even if nothing more had been accomplished there, what existed in 1972 would suffice to make the complex one of the most important in the industry.

Though impressive enough when seen at ground level, the complex should be viewed from one of Dallas's downtown skyscrapers to obtain a full appreciation of the magnitude of the accomplishment. From there one can see most of the old Trinity Plain, now occupied by the string of marts that make up the Market Center. It amounts to a new city, a nearby expansion of Dallas made possible by the Stemmonses and Crow. There literally is nothing else like it in the world. The center is an audacious concept. Had all the structures been proposed in the early 1950s, the center would have seemed fantastic and incapable of being constructed, much less occupied successfully. Step by step, building by building, Trammell Crow had made the fantasy real and peopled it.

Other marts would be built later on, and even now Crow is planning additions to what is the largest market complex in the world. Occasionally he is asked when it will be finished, and he replies that he will be finished before it is. Crow still regards the Market Center, in its entirety, as his greatest achievement in the area of development. Writing about the complex in 1981, he said:

Taken together, the Market Center and hotels are a city in themselves, a commercial city, which some have called a new downtown for Dallas. It works. The Trammell Crow Company's competitors still come to observe what we do, and emulate our ways.

The Center will endure—and by that I mean not simply exist, but be important, useful, and beautiful, We made our mark on other parts of the city as well, refashioning the skyline in the process.

But there were in this period plans for other projects taking shape elsewhere, and they merit description and analysis before we return to further developments on the Trinity Plain.

FOUR

Beyond Dallas

You have to present Americans with products and services they see as familiar, yet as new and wonderful. Know what the market will take, will pay for, will accept. Then study the situation and provide what is required. It is as simple and as complex as that. It is what was behind the successes of most businessmen, and in our century could be perceived in the careers of such diverse individuals as Henry Ford and Ray Kroc, and such products and services as cable television and frozen foods.

—TRAMMELL CROW, 1987

While usually appearing relaxed and confident, Crow is in reality a nervous, jittery person who has always found it difficult to concentrate on any single project to the exclusion of all others for more than a few days, and in some cases hours. He dislikes formal meetings, sometimes appearing late and walking out to conduct other business while the participants go over thorny and complex matters. Crow has little interest in the past, a good deal of interest in the present, and less in the distant future, unless attempting to forecast demands for commercial and residential projects.

He is very good at tactics but doesn't really care for strategic planning. More than anything else, he enjoys "working deals."

One of Crow's central business principles is to be involved with several deals at the same time so as to keep busy and constantly challenged in different ways. He also likes to range all over the map (literally and figuratively), pursuing widely varying objectives and finding changes of scenery stimulating. In 1958 Crow had already put up the Decorative Center, the first phase of the Homefurnishings Mart, and was completing work on its second phase. The Trade Mart's first phase was to be concluded the following year.

At the same time as the warehouses and marts were going up in the Trinity Plain in partnership with the Stemmonses, Crow was engaged in operations with other parties elsewhere in the city. He put up a garage for the Dallas Federal Savings & Loan in downtown Dallas. Crowlocke, a company organized by Crow and Eugene Locke, developed parking lots and garages in the area. It also constructed the Hartford Insurance Company Building, a 14-story, federal-style edifice that Crow helped design and that to some seemed out of place in Dallas. Crow thought it suited Hartford's character and image, and would attract tenants. The Hartford Building was his first in downtown Dallas, and many others would follow. To a large extent, the city's skyline today bears Crow's strong signature.

The Hartford Building was a success, drawing tenants without much difficulty. Its tenants were drawn partly by its location and the penchant Dallas businesspeople have for embracing the new. There also was the matter of rents and services: As was his custom, Crow was competitive in the former and offered more of the latter. When the building was completed in 1959, Homer Rogers and Crow moved their headquarters there from the Doggett Building, taking a penthouse office that was a far cry from the cramped and old-fashioned quarters they originally occupied.

Crow's interest in architecture is reflected in all his important buildings, including the Hartford. His style is eclectic; there is no single building that could be said to epitomize his taste. In his travels Crow would see buildings, or parts of buildings, that were attractive to him, and in time, when appropriate, he would incorporate the admired features in one or another of his structures. The only major criteria other than comfort and beauty were appropriateness and efficiency. Crow considered matters of cost very carefully. He was perfectly willing to provide luxury, but only when convinced he could get the proper prices for it in rentals.

No sooner was the Hartford Building completed than Crow started work on an office building complex with the Stemmons brothers, not far from the Market Center on the freeway. They decided to call it the Stemmons Towers. It was a four-building, $4.5-million complex that Crow constructed in order to stimulate high-rise activity near the Trade Mart. The Towers had ample parking in front, a sculpture garden, and a reflecting basin. Not particularly well accepted at the time, today the Towers is seen as a fine example of postwar office architecture.

Initially Crow worked with architect Harwell Harris, who later remarked that while he "wasn't a graduate student when I worked for Trammell," he could accept some of his ideas, but "expected that most of what I proposed would get built." As with so many of his projects, Crow found himself in a financial bind that required him to make changes, one of which was his architect. The economical Harold Berry, who came in toward the end of the Towers construction, did shave costs.

When the second building was completed in 1964, Crow moved his office there. But he still used the furniture that had been in the Doggett Building; not until the move into 2001 Bryan in 1974 would Crow relent and purchase new furnishings. Furthermore, the office layout remained the same—his desk was still in the open, with no forbidding walls, no place to hide.

Rentals at the Stemmons Towers came slowly at first. Crow had a hunch the offices would suit some of the foreign-owned operations newly arrived in America, and he set out to sign them on. It took him four years to convince one British firm to lease office space in the Towers, and at that he had been obliged to provide a year's free rent. Cooper thought their reluctance was because while Dallas had become a major metropolis and was one of the fastest-growing and most affluent American cities, foreigners found it difficult to take the city seriously, accustomed as they were to dealings with older coastal and midwestern centers. A world trade mart would increase the city's stature and overcome such hesitation. Cooper knew its presence could also give the Towers additional credibility, creating a pleasing and profitable synergy.

While Cooper pressed for the mart, Crow was busy elsewhere. He had started putting up warehouses in other parts of the country during the early 1950s, with Denver and Atlanta the first stops. His earliest clients were businesspeople and companies like Mueller Brass, with whom he did business in Dallas and who also had interests elsewhere.

Atlanta was in the midst of the same kind of real estate boom Dallas was experiencing. In 1959 the city issued more than $113 million in building permits, and for the first time sales of the top 40 firms went over the $100 million mark. Office buildings and hotels were rising in the downtown, expressways were under construction, and malls and shopping centers were sprouting in the suburbs.

Crow did not have to blaze a trail in Atlanta. Leo Corrigan had preceded him, constructing the 25-story Fulton National Bank and other structures in the downtown area. In fact, in the late 1950s and early 1960s developers from all parts of the nation were entering the Atlanta market. In this regard, Crow was something of a latecomer. Moreover, in the beginning he was not considered a major factor and was not particularly well known, as exemplified by the constant misspelling of his name as "Trammel" in the local trade press.

Crow immediately and instinctively realized that to compete successfully in Atlanta—or any place other than Dallas—he would need an on-site partner who knew the market and had the requisite contacts for success. At this time Crow was going from deal to deal, taking those opportunities that came his way, improvising daily, and was learning that others, more knowledgeable regarding their markets than he, would have to be brought into those deals. In time Crow would have national, and then international interests, and wherever feasible he operated through a local partner or native who understood and accepted his approach to business.

In Atlanta Crow worked through Frank Carter, the agent who had found the location for Mueller Brass's warehouse, and his friend and associate, Ewell Pope. They operated a brokerage firm known as Pope & Carter, which soon was doing most of Crow's work there. Out of this in 1968 came the organization of Crow, Pope & Carter, which concentrated on residential development. For a while it was Crow's flagship firm in Atlanta, handling all of his development operations in that city and later on elsewhere as well. By the late 1960s it seemed that Crow, Pope & Carter might become a national entity, as well as a leader in the residential field in the Southeast.

Residential real estate development is quite different from commercial development. The usual procedure in commercial real estate is to develop a concept, either after being approached by a potential client or investigating to see if there is a need for a project that renters might appreciate. Then the developer obtains financial backing, buys the land, hires the architect, locates a builder, and in other ways sets things into motion.

Jefferson Brim Crow, father of Trammell Crow.

Mary Simonton Crow, mother of Trammell Crow.

Three of the eight Crow children: Stuart (age 4), Trammell (age 3), and Davis (age 5), pictured in 1917.

Trammell Crow as a seventeen-year-old high school junior (1931).

Trammell Crow and his bride, Margaret Doggett (August 14, 1942).

The Crow clan today. Family members are identified on page 254.

FINE HOMES FOR BUSINESS FIRMS

Let us build YOURS *in Trinity Industrial Park*

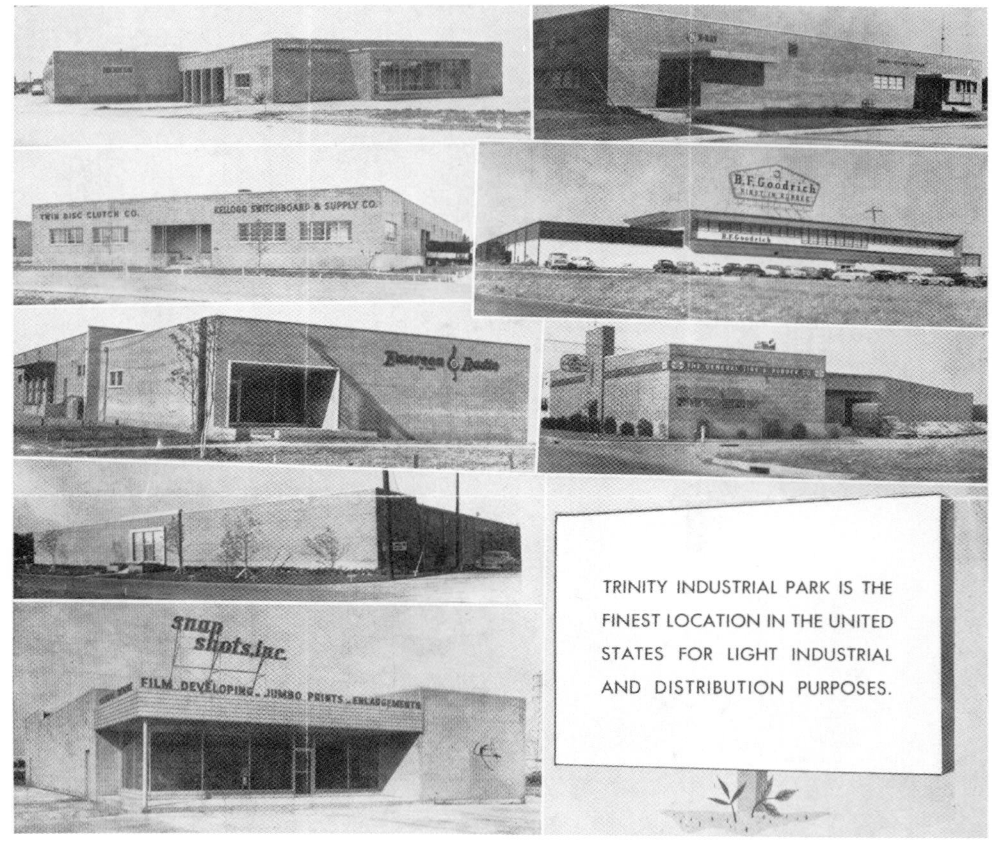

Here are a few of the buildings we have provided for business firms during the last few months. We provide site, planning, construction, financing — one package. Your inquiry is invited, either direct or through your real estate agent.

TRINITY INDUSTRIAL PARK

PART OF THE TRINITY INDUSTRIAL DISTRICT

TRAMMELL CROW — OWNER & DEVELOPER

425 S. FIELD STREET TELEPHONE RI-5847

Advertising flyer depicting several of the earliest Trammell Crow warehouses (1956).

Burial of a time capsule at the groundbreaking ceremony for the Dallas Furniture Mart (1957). Pictured: B. F. McLain, President, Dallas Chamber of Commerce; Trammell Crow; R. L. Thornton; George McNeff, Falcon Manufacturing Company.

Signing of loan agreement by Equitable Life Insurance Company for development of the Dallas Trade Mart. Top: Storey Stemmons; Trammell Crow; John Stemmons. Bottom: L. G. Raney, Vice President, Equitable Life; J. D. Francis, President, Mercantile Bank (1960).

John Stemmons and Trammell Crow in the atrium of the newly-opened Apparel Mart (1962).

Atrium of the Dallas Trade Mart, November 22, 1963. At the podium, Trammell Crow has just announced that President John F. Kennedy has been shot en route to the reception.

Visiting a development site in Atlanta: Ewell Pope, Trammell Crow, and Frank Carter (1968).

Ground-breaking ceremony for San Francisco's Embarcadero Center (July 1, 1968). Left to right: Mayor Joseph Alioto; Trammell Crow; David Rockefeller; Justin Herman, head of the San Francisco Redevelopment Agency; and John Portman (holding shovel).

Bill Cooper and Trammell Crow, at the time of their trip to China (1978).

Residential real estate involves a different sequence of problems and possibilities. For one thing, land isn't acquired as step three or four; often it is the first thing one thinks about, and not necessarily with a mind toward immediate construction. Rather, the parcels go into an inventory, awaiting the moment it makes good commercial sense to start construction. Builders can't wait until the demand is felt in a locale, since by then the price of desirable real estate is sky-high. So they are always seeking locations that might make them look like geniuses four or five years down the road.

Then too, construction begins with an idea of who the developer hopes will rent, but with no actual customers in sight. In a way this beginning is akin to what Crow did when constructing some of the early warehouses and the Homefurnishings Mart, in that he had no renters initially. The difference was that Crow knew whom he wanted as renters: those Dallas businesses and the major Carolina furniture manufacturers. In addition, he had a good idea of their needs, the competition, and just how much he could charge. No such well-defined market existed in the rental of, say, garden apartments. It is one thing to know that shortages exist and that newspaper advertisements and discussions with real estate agents provide a ballpark idea about what prices can be obtained. Developers don't know when they seek an architect what the circumstances will be when they are prepared to rent or sell or how their designs will appeal to potential renters and buyers.

Projecting housing needs is a chancy operation. Suppose it seems there will be a shortage in a year or so. Perhaps Congress passes a new bill encouraging housing construction. These might be signs for aggressive individuals to start moving. The trouble is that many get the idea at the same time, and the result is overbuilding and distress.

Hundreds of developers have come up short in every real estate bust, stranded with inventories of land that have to be carried at high interest rates, and unsold and unrented houses and apartments that often have to be dumped at distress prices. Whether or not Crow thought at the time that Crow, Pope & Carter could avoid the pitfalls of residential speculation, that firm's experience was to prove no exception to the rule.

Warehouses remained the core of the Crow operations, and their success prompted him to seek ways of supplementing and supporting them. To accomplish this Crow decided to organize a company which would operate a public warehousing business. A public warehouse receives goods from others, stores them and ships them out as directed by that owner, usually the company which manufactured the goods.

The purpose and service of this function is to permit manufacturers to ship a large volume of goods to a warehouse located conveniently within the region where the goods would be ultimately sold, and then reshipped in smaller quantities to the end customer. The cost savings and service provided to the ultimate buyer are obvious.

By having his own public warehousing company Crow would expand his own warehouse developing operation, would be able to produce income from vacant warehouses which showed up from time to time, develop rental properties through the public warehouse company's clientele, and all in all, become a more fully integrated warehouse developing operation.

The operation called for the hiring and retaining of a custodial staff and other support personnel, which were not required in the other warehousing efforts. In 1969 the venture was spun off as Trammell Crow Distribution Corporation (TCDC). The company started out with facilities only in Denver, but within a few years was in 13 cities. Later on trucking operations were added, and TCDC became a major factor in the field.

In the late 1960s, however, residential development in all of its phases seemed more promising even than warehousing, and the affiliation with Carter and Pope intrigued Crow. These two partners reminded Crow of his relationship with Homer Rogers. Like Rogers, Carter was conservative and income-oriented. Crow considered Pope an absolute genius at selecting properties, but this talent was coupled with what he called "proclivities to financial impecunities and a tendency to commit himself and his partners beyond likely financial expectancy, and beyond his own and his partners' pre-arranged limitations." In other words, Pope was an imaginative high roller who when certain of himself would roar ahead regardless of costs. Carter once remarked that the difference between them was that they both liked to play poker, but that Pope would go into the game with a fat wallet, while he took along only what he thought he could afford to lose. Crow thought that perhaps it made sense for two such people to work together, each providing balance for the other.

Crow's Atlanta projects performed outstandingly in the late 1960s and early 1970s. Due largely to Pope's boundless energy, Crow, Pope & Carter expanded into other places as well—Florida and North Carolina in particular—usually but not always with local partners. By then, however, troubles had developed in the relationship between Pope and Carter. Carter was growing increasingly nervous about Pope's plunge into real estate. Even more than Crow, Pope craved growth, accumulating more raw land for future development than Carter thought the markets could handle. It wasn't prudent, but of course prudence had never been one of Pope's hallmarks. On his side, Pope was increasingly irritated at Carter's apprehensions. Carter and Pope remained friends, but each wanted an enterprise that better reflected his personality and proclivities. Clearly the time had come for the two men to part, and they did so in 1971.

Crow perceived fine qualities in both men and was troubled by their "divorce." "You may want to break up," he told them, "but I want to be partners with each of you." The answer was to form two new entities, one with Carter, the other with Pope. The new Crow, Carter & Associates began with several major regional shopping malls in Atlanta, and then went on to build malls elsewhere, starting in Charlotte, North Carolina, in addition to erecting office buildings. Crow, Pope & Land (the last named being A. J. Land, who became the company's operating partner) was concerned primarily with apartment houses and hotels.

Perhaps it was inevitable that Carter would want to go it alone after a while, and a few years later he bought out Crow's interest, this break-up also occurring in an amicable fashion. Crow retained a share in some joint pursuits, but all new development came out of what now became Carter & Associates. The Crow, Pope, and Land Association continued until 1976, developing properties in many parts of the United States.

Crow's work in Atlanta brought him into contact with John Portman, one of the great architect-developers, with whom Crow had a relatively short, rewarding, and occasionally stormy relationship. The partnership was one of two strong egos, each concerned with quality and performance, capable of working together on some projects, but bound to clash and eventually separate.

Crow called Portman in Atlanta in 1960, when Crow was working on the Atlanta Decorative Arts Center. The two had many common interests. Crow thought it sensible that they compare notes, which they did. He perceived in Portman a talented man who was not comfortable with conventional structures, which is to say they were quite similar. From the first it was clear that Portman and Crow were on the same wavelength, arriving at similar conclusions from different perspectives. Portman was an architect by training, the leader of the firm of Edwards & Portman. He chafed at designing buildings for others and yearned to develop them himself, and thus he became a developer. In contrast, Crow had started out as a builder and developer, and as has been seen, went from there to cultivate an interest in architecture.

At the time most of Portman's projects were done with Ben S. Massell, Atlanta's largest and best-known developer, and knew that as long as the relationship existed, he would have to hew to Massell's wishes. With Massell's backing, Portman had created a Merchandise Mart in a vacant, multistoried warehouse building close to the city's center. Portman, who became president of that Mart, and chaffed at what he considered Massell's imperious ways. He thought it might be different with Crow.

Crow and Portman united to put up the rather small project, the Decorative Arts Center, which went well. Soon after they decided to construct a new building to better house and expand that Mart. It was to be a 20 story, 880,000 square-foot building to be on Peachtree Street, located on what then was the fringe of the central business district. What Portman envisioned was similar to what Crow intended doing in the Trinity, though of course Portman wasn't building on fresh ground, but rather attempting to rejuvenate a depressed area. The choice of location was a bold move on his part, and it was his boldness that set him apart from the more conventional and conservative Atlanta developers of the period.

The first step was to purchase a site for the mart. They put together a package, interested others in providing some capital, and started work. Crow's main contribution was to provide the credit needed and arrange for financing from outside partners. He did not have much input into the construction, but did push for space for the inclusion of lamps, gifts, carpets, and apparel displays, which made sense, since without these merchandisers the mart would not have been fully leased. After the project was completed Crow and Portman bought out the other partners.

The Peachtree Center transformed downtown Atlanta. It not only provided an attractive and highly commercial environment, but resulted in a new linkage between the financial community and the younger, more imaginative real estate interests.

Soon additional projects were discussed, planned, and constructed. The $12-million Greenbriar Shopping Center, the first enclosed mall in Atlanta, followed. And there were other projects. None was as well-known or as influential as the Peachtree, however, which Crow and Portman envisaged as a complementary cluster of buildings to revitalize downtown Atlanta. They purchased a parcel of land just to the south of the mart on Peachtree Street and obtained a long-term lease on another. On these they planned to construct a modern, distinctive office building, which Edwards & Portman would design. The southern and northern segments of 230 Peachtree Street, each 30 feet wide, were connected to the rest of the building by land bridges. This gave the building an airy feeling, which juxtaposed nicely with the bulk of the neighboring Merchandise Mart.

Meanwhile on his own and without Crow, Portman was attempting to build a hotel in downtown Atlanta. He had employed a manager from a local hotel, and expected to complete the construction and then manage the facility himself.

Crow did not come in on this one. He was not critical of Portman's proposed that design, which he thought excellent. The hotel was to feature an atrium, 120 feet on each side, rising 22 floors, with expansive vistas, intriguing pockets of space, and interior glass elevators. Portman became famous for having introduced this feature, and it is true that he pioneered atriums in hotels. But the idea came from Crow's Trade Mart. Crow recalls Portman's visit to the Dallas mart: as he and his guest walked through a low-ceilinged entrance, the great cavernous hall exploded before them, and Portman was suitably impressed. The fact that Portman's atrium was inspired by Crow's does not discredit the Atlanta architect—after all, Crow also borrowed the idea.

Somehow he had gotten into financial difficulty on the project, and told Crow he would lose everything he had put into it unless he could raise the necessary funds from outside sources. Accordingly he asked Crow to try to secure the money and come into the deal with him. At the time Crow did not have the $500,000 required, but he thought he knew where it could be had. Crow had known Fritz Hawn, a Dallas Financier, for several years, and the two men had worked together on numerous deals. Indeed, Hawn had already come in with the two men on the Greenbriar Shopping Center, so he knew Portman as well. Now Crow approached Hawn to inquire whether he would invest in the hotel project.

Hawn showed little interest, so Crow then proposed that Hawn buy his and Portman's portion of the Greenbriar, in this way providing the funds Portman needed. Hawn agreed, and consequently Crow became a partner in the hotel.

While recognizing the viability of the hotel, Crow did not believe they should attempt to manage it. After some discussion they decided to try to sell it to an operator. Hilton turned them down, as did Marriott and others. Then Hyatt came into the picture and made an offer. Crow insisted they sell, while Portman held back. In the end it was sold to Hyatt, which at that time had only two or three small, non-central city holdings, and this became the launching of the Hyatt chain of today and of Portman's career as a premier designer of hotels. Crow now recognizes that he was wrong in this judgment, but the act was another demonstration of his basically conservative nature.

Crow and Portman had several other joint projects, the most important being a mart in Brussels, Belgium, which will be discussed in some detail in a subsequent chapter. But they were doing more away from one another than in tandem, and in addition Portman had to run matters at Edwards & Portman. Crow's contributions of cash were small, though he did arrange financing and signed notes. However, as time wore on he became more and more concerned with Portman's difficulty in meeting budgets. He was finally impelled to ask for a separation when Portman told him of plans for a second hotel in the Peachtree Center, the design for which Crow thought was poor, both financially and operationally. In addition, Portman wanted to put up a hotel in Los Angeles, and Crow held an even lower regard for that idea. So Portman went into those deals on his own. Both projects ran into financial problems, requiring additional funding and a tremendous spate of inflation in order to survive.

The two troubled hotel projects were followed, however, by Portman's venture in Detroit, the Renaissance Center, which was an artistic success, but a financial failure at the time. As the name indicates, the center was meant to both kindle and symbolize the renewal of Detroit's central city. It was unable to attract sufficient business initially. Henry Ford II, who was the principal backer of the project, was roundly criticized for failing to perceive the realities of the economic situation. There is something to be said for such creativity, Crow reflected later on, although he thought it should be tempered with a sense of realism.

What lessons might be derived from the successes Portman and Crow enjoyed together, and why did they part company? The successes were due to the fact that they had similar objectives, mutual respect, and complementary styles. Also, they needed one another. Portman had indisputable architectural talent, and Crow had experience in real estate development and access to backers.

By 1969, however, it was becoming clear that Portman considered Peachtree a sort of demonstration, a laboratory for a new kind of urban development. Crow had interests in a wider variety of ventures, and he harbored doubts about Portman's visions. There were smaller, but still abrasive, factors contributing to the dissolution of their relationship. Portman maintained his independent architectural practice, which was used for their joint projects. His fees were substantially higher than Crow was accustomed to paying, while Portman chafed at Crow's inability to understand that comprehensive designs such as he was involved with were by necessity costly. Years later, Crow wrote an essay on architecture that perhaps indicates his general distrust of its practitioners, and a criticism of what he perceived in Portman and some of the other architects with whom he had dealt during his career. He said of developers:

> *We are all, those of us who survive, market controlled. We learn what the market will take, what reality will allow, and go for the best possible. Perhaps some can afford qualities at the ultimate top of market realities, while others cannot. Some, as surely do we, yearn to construct the finest. But the arrogant architect who considers a commission a charter of rights to himself to "create"—not faithfully taking into account the judgments, needs and limitations of his client—is, in my view, a charlatan, to say the least. I am always reluctant to donate to public buildings where, due partly to the nature of committees, client control is hardly existent, and rarely competent.*

Recall that Crow credited Dallas with providing fertile soil for individuals such as himself after World War II, but there were other cities in which aggressive developers could and can do well, including Atlanta, Los Angeles, San Diego, and Denver. San Francisco was also on Crow's short list of appealing areas. While working with Portman, Carter, Pope, and others in Atlanta, Crow scouted San Francisco, which seemed on the verge of an important building boom. He felt that a clothing mart there could do for the West coast what the Apparel Mart had done for Dallas.

The San Francisco Redevelopment Agency had acquired near the waterfront 40 acres of land that once had been the site of a major produce market, intending to use the space for urban renewal. About two-thirds of the land had been earmarked for the Golden Gateway Center, which was planned as a residential project of some 2,500 units. Six or seven acres were to be parks, and the rest of the land was to be held in reserve. The agency seemed eager to have Crow participate in the undertaking and invited him to explore the area and come up with a plan.

M. Justin Herman, the director of the San Francisco Redevelopment Agency, took Crow on a tour of the site in 1966, showed him the parcel of five blocks, and asked if he would be interested in assisting with the redevelopment. The agency would provide interim construction financing if the parcel was kept intact. The land had been appraised at $30 a square foot, which is what they were to pay for it.

It was a clean and appealing deal. Crow accepted, and put $250,000 of his own money into the arrangement for starters, which indicated his confidence in the venture. Then he organized a team, comprised of Portman, David Rockefeller, James Caswell, and Cloyce Box. Caswell was an Atlanta developer who played a significant role in several of the projects Portman and Crow had put up in that city, while Box was an old friend, a former football player and later an attorney, who had entered the construction business at the George A. Fuller Company. Rockefeller provided equity financing, Portman the design, and Box the construction. Crow directed the operation, while Caswell might be considered the chief operating officer, for which he received a portion of Crow's and Portman's share. Metropolitan Life was the initial mortgage lender; Prudential Insurance provided the debt financing for a hotel and second office building on condition it would be a 50-percent partner in both. Prudential also insisted that Hyatt be brought in as operator of the hotel. Thus was born the Embarcadero Center, and the groundbreaking took place on July 1, 1968. It was the biggest deal Crow had undertaken to that time—$300 million, of which he would have an 8⅓ percent interest. Money was fairly tight at the time, and all involved seemed pleased with the deal and delighted that the way ahead seemed clear.

Construction on the first office building began soon after groundbreaking, the plan being to bring it in by late 1970. The building was completed in early 1971, and the first tenants moved in soon after. Part of the agreement with the Redevelopment Agency was to start construction on the hotel at that time, which is what was done, and it was ready for opening in the spring of 1973, on schedule and on budget.

An incident involving that hotel, which was to become the San Francisco Hyatt Regency, illustrates problems developers often have to face.

Large cost overruns developed as the project neared its conclusion. The Prudential agent expressed a willingness to increase the loan to cover this, but asked in return for a greater interest in the project. A solution was reached in which the insurance company got an 80-20 split for the first four years, after which it would go to 50-50, but only after Crow's attorney, Joe Stalcup, aggressively encountered the Prudential representative in such a manner as to impair Crow's long standing relationship with the company.

During this period the partners negotiated with Levi Straus for a major lease in the second and third office buildings. The apparel company wanted 200,000 square feet in a configuration that required some replanning. The second building was completed in 1974. At that time the partners had some disagreement with the Prudential as to whether they should continue on or not, due to unsettled economic and financial conditions (which are detailed in Chapter 7). Prudential exercised its option to withdraw from the project and not become a partner in the third building. David Rockefeller stepped into the breach with the needed funds, and construction began in 1975.

When Embarcadero 3 was completed, the partners were able to lease only around 10 percent of its space, and for a while the financial situation was quite precarious. However, the market turned around soon after, and within less than a year and a half the building was 90 percent occupied. The Embarcadero eventually became enormously successful; Crow could see it would be so almost from the first, and encouraged further expansion. Crow always had confidence in the project and hated to sell, but in 1974 he was obliged to dispose of everything he could in order to remain afloat, and so he sold his share to Rockefeller.

By then, Crow came to feel that he could no longer risk underwriting his partnership with Portman because of the ever present problem of cost overruns. He learned from Portman about architecture and had given the architect the background he required in development. The two men had come together out of mutual need and respect. After the Peachtree project they went on to create the Park Central Office Park development in north Dallas and other projects, but their relationship came to an end in 1974.

Given Crow's industry position at the time, one might have expected that the operation had expanded to the point of having layers of management, organizational charts, lines of command, and all the other paraphernalia associated with big business. This was not so with Crow's operations—which still could not correctly be called a company. And it certainly was not a single corporation. Rather, Crow's operations were a series of unrelated deals, each with its own terms, partners, financing, potential, and problems. Some deals were quite small—a warehouse in Dallas, for example—while others, like the Embarcadero and Peachtree Center, were among the nation's more significant projects.

There were certainly no trappings of big management to hint at the scope and significance of the operations. Crow's offices at the Hartford and then at Stemmons were more elaborate than that bare room at the Doggett Building, but they hardly would have impressed visitors—which wasn't a problem, since Crow had few at the time. Headquarters really was wherever Crow happened to be. He kept track of the various deals through memory, and the scarcity of written contracts presented bewildering difficulties for his associates later.

Those who dealt with Crow regularly became accustomed to the way he did business, but he always retained the ability to amaze. There was the time when mortgage broker Glenn Justice successfully negotiated a $1-million mortgage commitment for a warehouse project, for which his fee would be $10,000. As usual, Crow was a little short at the time, so he scrawled an I.O.U. for that sum on a sheet of paper, handed it over, and said, "Now Glenn, put this in your desk drawer and I'll pick it up in a while." Two months later he dropped by Justice's office with the $10,000 check.

As time passed, these casual business methods became unsatisfactory. The approach and techniques fitting for a Dallas warehouse were not appropriate for an office complex in Atlanta. Geographically and functionally, Crow was spreading himself thin. As has been seen, Crow was perfectly willing to permit partners to take responsibility for operations while he went on to new projects; by the late 1960s he had to contemplate delegation of executive responsibilities as well.

What was the alternative at that time? He might have put together an organization made up of administrators, associates, and assistants, but such structure was alien to Crow's nature. He had never headed an organization such as this, not even in the Navy. Crow was a free-wheeling man who was used to working with others of like mind. There was no pressure at headquarters for change—largely because, as noted, there was no headquarters in the conventional sense of the term. At Crow there were no divisions such as one finds in all large companies, and no such elements as middle management and hierarchies of power and responsibility or line and staff distinctions.

Staff executives have a stake in structure and usually can be counted on to seek to centralize power, while their counterparts on the line often support delegation of power to their units. In companies with strongly delineated management/staff levels there may be a creative tension out of which evolve responses to challenges. This tension was not present at the various Crow interests, where each of the partners had a great deal of autonomy, there was no staff, and the central figure had a positive distaste for bureaucracy. A Crow partner in Atlanta, for example, may not have known anything about operations in Dallas or elsewhere, or have seen any reason why he should.

Crow believed every company with a bureaucracy had a tendency to ossify, resulting in stifling of creativity. He was determined that ossification would not occur in his operations. He wanted very much to keep the business simple, direct, and as personal as possible. Crow still answered his own telephone, invited all around him to read his correspondence, and encouraged an unfettered, playful attitude toward the business.

In his 1969 book *The Peter Principle,* Lawrence Peter admonishes businesspeople to be wary of pitfalls of entrenched power groups. "In a hierarchy, every employee tends to rise to his level of incompetence (the cream rises until it sours)," he wrote. "In time every post tends to be occupied by an employee who is incompetent to carry out its duties." Rationalizing his preferences, Crow considered that Peter's warnings were good enough reason to continue on as he had in the past. Instead of a hierarchy, he had several quasi-independent, hard-working and intelligent partners, each doing whatever was required to ensure the success of the particular venture with which he was involved.

So Crow seemed to have evolved a method of operation that suited his business and personality ideally. All was going well. But then Crow, rarely a strategic thinker, made a major blunder and learned an important lesson. Those were intoxicating times; everything seemed to be falling into place for Crow, who seemed incapable of making a false move. He entered a new area of interest, construction, at a time when it seemed most attractive. As it turned out, he hadn't done his homework on the industry and company involved.

In his classic study of General Motors, *Concept of the Corporation*, Peter F. Drucker wrote of innocently asking the auto company's board what business it was in. The question set off a discussion and debate out of which came major alterations in the firm's strategies and organization. It is the kind of question businesspeople should always ask themselves. Drucker later was quite skeptical about conglomerates—firms that claimed they could create synergy between disparate units—and even questioned some horizontal and vertical mergers.

If in this period someone had asked Crow what business he was in, he would have answered, unhesitantly, "Real estate development." From warehouses Crow had gone to exhibition buildings to marts to office buildings, parking lots, hotels, malls—all involved with real estate development. Those first warehouses went up in 1948; less than two decades later he was one of the nation's largest real estate developers. It was heady stuff.

In the early 1960s Crow felt confident enough to engage in two efforts at expanding beyond development. These forays seemed to be sensible extensions of existing business.

The first expansion involved providing assistance in the creation of a company that would aid in his financing. As always, money was the key. Crow constantly needed additional financing because of his rapid growth. Recognizing that to a certain extent he was at the mercy of his bankers, and aware of advantages which might flow from a controlled pool of investment funds, in late 1959 Crow took the classic step of integrating backward by organizing, with his friend Eugene Locke, a company which would become a major source of his financings. In March of the following year that entity, Wallace Properties, Inc., was incorporated in Delaware. It was headed by E. E. Wallace, Jr., a former senior vice-president at Republic National Bank who had handled several financings for Crow.

Under the terms of its charter, Wallace was to have a wide range of interests, all of which involved real estate. The company could erect structures, manage them, engage in real estate development, and, most important, make interim construction and mortgage loans. Several Texas businesspeople were involved in Wallace at its beginning, providing seed capital and, in the case of Crow, properties in exchange for stock. In return for several of his holdings, including the Doggett Building, Crow received an

11.2-percent interest in Wallace. Among the others at Wallace were three of his old friends, Henry Beck, a local contractor, Tom Lively of Centex Construction, and Eugene Locke. Involved in a minor or peripheral way in Wallace were Eddie Kahn and Henri Bromberg, who contributed their shares in Crow-managed warehouses in the Trinity Industrial District. Some Crowlocke holdings also went into Wallace. So the company started out as an affiliation of friends and associates with several office and office-warehouse buildings, garages, and apartment houses in Dallas, San Antonio, and Shreveport, along with undeveloped land in Los Angeles.

Soon Wallace expanded beyond its original form. In January, 1961, it acquired Gibraltar Investment Company of Los Angeles through a stock swap. Gibraltar's primary asset was Institutional Mortgage Company (IMC), a major investor in Federal Housing Administration and Veterans Administration mortgages. Wallace intended using IMC as a vehicle for acquiring mortgages that would then be placed with institutional investors. As the firm was evolving into something far more ambitious than it had been initially, Locke assumed the chairmanship, while Wallace served as president. Later in 1961 the company changed its name to Wallace Investments, which was a more accurate description of its activities, and established several subsidiary units, one of which was Wallace Realty Mortgage.

Wallace was an example of Crow entering an area of opportunity at the right time. Real estate was booming, and Wallace rode the crest of the wave. From the first Crow was one of its major customers, although soon Wallace was lending to a wide variety of others as well.

In 1963 Realty Mortgage acquired and then assumed the name of Connecticut-based Lomas & Nettleton, one of the nation's oldest mortgage bankers, which also had interests in insurance. Admiral Fire Insurance of Houston was also acquired. These acquisitions signaled Lomas & Nettleton's intention of becoming a full-line financial company. After several management shuffles and additional acquisitions, Jess Hay, a partner at Eugene Locke's law firm, took over as president and CEO.

In all, Lomas & Nettleton was a successful effort that provided Crow with important support at a time of necessity. Not so successful was his attempt to expand into what at the time seemed an obvious area directly related to his business—construction.

By the mid-1960s a portion of Crow's construction work was being done by George A. Fuller, and his relations with Cloyce Box were quite close.

The prestigious Fuller firm had been founded by its namesake, the dean of M.I.T.'s Engineering School. Working directly with Andrew Carnegie, Fuller put up some of the nation's first steel frame buildings. The company refined high-rise techniques and was responsible for such landmark edifices as the Tacoma Building in Chicago, New York's Flatiron and Times Square buildings, and, after World War II, the United Nations complex and the Lever House.

Fuller, the world's largest office building construction contractor, was highly regarded in the industry. It had the reputation of never having had a losing year and of maintaining a strong balance sheet. Despite its reputation, George A. Fuller's stock, listed on the New York Stock Exchange, was selling for less than its break-up value, and so was perceived as being underpriced.

The reasons for being underpriced were clear. Construction was a cyclical business, and even firms like Fuller could have bad spells. Yet how could one explain that unbroken skein of profitable years? Part of the reason for that record was manipulation of figures. The company reported earnings on a percentage-of-completion basis, which meant that income could be delayed or carried forward simply by underreporting or overreporting the percentage of completion.

Crow was aware that Fuller could not be as miraculously healthy as it seemed on the surface, but he also determined that the firm could be purchased rather inexpensively, preferably, as is his wont, with borrowed funds. Fuller would make a wonderful captive concern that could perform most of his construction. It was all there—integration, synergy—all the current buzz words used to describe forward-looking, progressive companies. Only Crow wasn't concerned with style and razzle-dazzle. All he wanted was a top-notch firm to do his construction. He was not trying to build an empire; he was only making what at the time seemed a logical move into a related field.

Crow once remarked that two impulses probably strike every developer at one time or another: to enter construction and to build and manage hotels. Crow's views on hotels will be discussed in detail later on. Here we will see why he felt the way he did about construction companies.

The smartest, most prudent developers stop before it is too late, considering that such an acquisition or diversification can bring more headaches than benefits. All think they can avoid the pitfalls. None do, not even Crow.

Indeed, his only previous experience in the construction field had not been successful, and so he should have known better. In 1963 he had invested in a small start-up firm in Dallas, the Coker Brothers Construction Company, which from the first had financial troubles while attempting to make a go of it in prefabricated housing. Coker had contracts to construct several apartment projects and to do some rehabilitation work. There were cost overruns. Crow thought he understood the business well enough to avoid such problems at an operation of his own, but he was wrong.

At the time Crow became interested in Fuller, Cloyce Box was its president, without much of an equity stake. Crow suggested that the two of them buy enough shares in the company to take command. It wasn't long before they decided to take complete control, and then take the company private.

Without appreciating the significance of the move, Box and Crow in 1965 arranged a leveraged buyout, the kind that would make news on Wall Street 20 years later when practiced by men like T. Boone Pickens and Carl Icahn. The buyout didn't attract much attention at the time. Despite its prestige and industry position, Fuller was a small company, and the nation's financial and business writers had more on their minds than this relatively isolated arrangement.

Crow employed the law firm of Sullivan & Cromwell on this deal. The attorneys agreed that such a transaction could be cobbled together, but they doubted, at least initially, that with all the involved parties, the innumerable contracts extant, the pension funds, the inevitable unhappy minority shareholders, and so forth, it would ever come to pass. But Box and Crow were sufficiently enamored with the idea of owning and operating this historic company and becoming construction moguls that they went ahead anyway.

Lou Crandall, a prominent figure in the New York financial community, was chairman and CEO of Fuller. His son-in-law, Bill Lawson, was a vice-president, and though they would not have selected him for the position, the circumstances were such that Crow included Lawson as a buyer/partner. Maurice Moore came in as an investor; he and Crow had put up some warehouses together in Dallas and elsewhere, and they had a comfortable relationship.

Crandall was a key factor in the proposal. With his strong manner and dominant personality, he literally took over the arrangements. Box could represent the purchasing group's interests, but having worked with and for

Crandall for several years and being relatively young, he was not wholly free from the chairman's ministrations. Crandall carefully acted to procure the best price for the shareholders, of which he was one of the more substantial, and treated Crow and Box like neophytes, which is what they were in this kind of business.

The Crow group paid between $15 million and $16 million for the company. They raised the money from some London insurance companies associated with Lloyds; Crow's direct investment was very small. As in all leveraged buyouts, they had swapped debt for equity, and thus owned a company with a much greater debt service than was usual for the time. The debt load didn't cause Crow undue concern, since that was the way all his real estate transactions had been structured.

There was a clause in the loan agreement that seemed minor but was to prove consequential. Crow gave a "net worth" guarantee to the banks, meaning that if the company's net worth fell below a specified level, Crow would be responsible for coming up with additional funds. In effect, he had guaranteed Fuller's success. He scarcely considered the possibility of such a dip in value, however. After all, the company had always been profitable, and Crow believed it would continue to be.

The closing was attended by dozens of lawyers and principals and was so complicated that it had a master of ceremonies, a three-hour rehearsal the previous night, and required 10 hours to execute on closing day. When all was signed, Crow and Box had that fine construction company, plus a pile of debts. With this, Crow entered the contracting business.

His operations since 1948 had brought Crow into contact with many general contractors, always as a client. Crow watched their work carefully, and after a while thought he had learned just about all there was to know. This had always been his way; by observing, questioning, and appraising he had developed skills in architecture, and he saw no reason why contracting should be any more complicated. With several major buildings under construction in New York and elsewhere, with two or three complex construction jobs completed but not closed out with the owners and the owners' architects, Crow found that he had purchased not only a construction company, but various relationships, some of which turned out to be adversarial. The complexities were vexing—and potentially dangerous.

Crow believes there probably is no business in which one takes so many risks for so few prospective rewards as in construction. Nor is there any business in which most parties to the undertaking attempt to lay off blame—when there are problems—upon the construction contractor. Consider what happens when there are errors in architects' plans: Many of them ascribe their own errors to the contractors. Fuller had these and other difficulties.

Crow and his associates also suffered from the continued employment of Crandall, which under terms of the agreement they had to accept. Crandall was more concerned with the well-being of his New York clients and connections than with The Fuller Company and Crow. Whatever interests he had in the company revolved around those of Bill Lawson, who turned out to be inadequate for his tasks, and who was resented by those who knew the only reason he held a vice-presidency was Crandall's patronage.

After a while Box managed to wrest power from Crandall, but rather than move to New York he remained in Dallas, as did Moore and Crow. One more mistake. Others were to come.

It was clear that things were going sour at Fuller, but those who should have stepped in were too concerned with other projects to devote sufficient time to the company. This benign neglect was typical of Crow. Ever optimistic, he would stay with a project as long as he could convince himself that success not only was possible, but would soon come into view. Fuller might have worked out well—in time. But Crow had no intention of devoting years to such an endeavor, not when there were disappointments almost every day, and when there were so many other entrancing projects beckoning, not only in the United States but overseas.

He could not simply abandon Fuller, however. There was that net worth agreement, which if conditions continued to deteriorate would require him to come up with substantial cash infusions; and Crow simply did not possess that kind of money. He might have to liquidate many of his holdings, and under the worst circumstances become illiquid and resort to protection under the bankruptcy laws. "If I lose everything, what would I do?" he asked a friend rhetorically, for he already knew the answer. "I'd have to start over again."

In 1968 the Northrop Corporation, which in addition to its aerospace enterprises was engaged in a number of construction activities around the world, indicated an interest in buying Fuller. Box and Crow were anxious to enter into discussions, as was Moore. Box and Northrop hammered out a deal that would enable the partners to break even. But Moore claimed that he should receive a larger share of the proceeds, leaving Box and Crow to take a loss. Tough negotiations followed. Somehow Box managed to close the deal, carrying Moore down to the last second. Crow's continued ownership of Fuller, with one crippled partner and another antagonistic one, could have led to real trouble. Box saved the day—even the year—on that transaction.

Looking back at the episode, Crow concluded that, given his temperament, the situation might have been salvaged had he done at Fuller what was being done with his development projects—if he had selected some bright, experienced person, given him an equity stake, and set him free. Yet there surely was more to the matter than lack of leadership. In taking over at Fuller, Crow was incorporating into his business a firmly established concern with an entrenched bureaucracy and a ritualistic method of conducting business, complete with organizational charts, lines of command, and all the corporate panoply lectured about in schools of business administration. Crow did not follow corporate mores. Over the years he had developed his own way of conducting business, of organizing, motivating, and proceeding. His methods resulted from his experiences, worked out through trial and error, from his observations of motivation, human nature, and from his hunches about what might work. Everything that went through his mind was gathered together and synthesized in an unsystematic but not unproductive fashion. His way of working was what Crow felt he needed at the time, and it certainly wasn't what existed at Fuller. So he moved on, incorporating the lessons learned from this experience into the patchwork quilt of concepts to be used in future deals.

FIVE

People

> *Make sure that your associates are partners, and not employees. That way they will always be working for you—or perhaps it would be better to say, with you. They cannot succeed if you don't, and so they will labor to make you look good, and do good. I want to be surrounded by successful, hungry, and wealthy people.*
>
> —TRAMMELL CROW, 1979

By the mid-1960s Trammell Crow's operations had expanded to the point where they might well have required some of the structural appurtenances of a more formally organized company, including formal recruiting and the establishment of career paths for newcomers. After all, Crow's varied interests were a big business by then, and big businesses were supposed to evolve into managerial hierarchies. But instead Crow simply continued conducting business the way he had been since the late 1950s, which is to say he went from deal to deal, accumulating assets and liabilities and partners.

Especially partners. The accumulation of partners was closely linked to Crow's strongly held notions about motivating workers. He believed his projects would thrive if those who managed them were partners rather than employees. That way they would always be working with him, and not for him, sharing risks and rewards, encouraged to do their best for the enterprise.

Crow did not develop his philosophy by consciously considering the factors of motivation, initiative, and work ethic. He was drawn to the idea of partnership by practical reasons: He had little in the way of resources, and what he did have, Crow intended to use to put up buildings, not to reward associates. Later on he would be portrayed as a brilliant theoretician in devising employee compensation, but his strategy was at the same time simpler and more complicated than that. Crow was an inspired improvisor who had the good fortune to operate at a time when the market rewarded those with his outlook, talents, and proclivities.

Crow thought the concept very uncomplicated: A person who is paid a salary won't do the same job as one who owns part of the deal. In discussing this approach, Crow remarked:

> *We get more than one hundred qualified applicants for every newcomer we take on, and have been singled out as one of the best companies to work for in the United States. Yet this really isn't quite accurate, since as indicated these people won't be working for us, but rather for themselves* with *us. When a good person is in that position you don't have to watch him too carefully, since he will knock himself out to make certain the project succeeds. Believe me, it does remarkable things to a person when he's an actual owner of a property he's working on. Become an owner, and things improve. I guess that's just the way people are created.*

His conclusion was based on experience, not theory, and was worked out in the context of doing business, not attempting to structure an enterprise. In creating his concept of partnerships Crow was pragmatic rather than ideological. He did not consult any body of management theory in attempting to resolve a problem. At the beginning there was just Trammell Crow. Then there were two people. Then there were five people. No one ever instructed Crow on how to get along with them or manage them.

Crow brought these people along as fast as they were capable of assimilating his approach to business, and then gave them an equity stake in an enterprise in which Crow himself was a partner—*partner,* not employer. One would assume he had to select these individuals carefully, after much study. Such was not the case in the beginning, when newcomers were taken on because of Crow's informed hunches about them. An instinctive leader, he considered what it would take to activate a person and then looked for those who seemed to have the kind of personality and the intelligence to be successful in the real estate development business. Later, when Crow became better known and reporters and columnists came to investigate his methods, they seemed surprised at how elementary the partnership concept was, and wondered why other firms didn't embrace it.

The approach to compensation evolved to meet changing needs and circumstances. At first Crow concluded the proper split would start at, say, 80–20, and then as the partners took on additional responsibilities, their stakes would rise to 50–50 and better. "There is something about this that carries fairness with it," he said, "the appearance of fairness if not actually fairness." His underlying belief was that if people are treated generously, they will return that generosity. "Fairness begets fairness, and loyalty begets loyalty, and generosity begets generosity. It's just the way humans live and work." Whether or not this causality actually existed in the beginning or grew over time is not certain, but Crow was convinced that by giving a great deal he was also receiving the same—devotion and dedication from hard-working, intelligent, capable people who understood that they would receive substantial rewards from their efforts.

Most of the Crow partners became millionaires, some very quickly and several many times over. Yet Crow often remarked that he didn't make his partners rich, but rather they made him rich. In practice it worked both ways. Nonetheless, his outlook inspired extraordinary gestures of appreciation. For example, in the late 1970s Crow and one of his senior partners, Terry Golden, traveled to Florida to inspect the work of several junior partners in converting some residential properties into condominiums. The project had turned out very successfully, and the partners were pleased with their share of the operation. On the plane taking them back to Dallas, Crow and Golden discussed the project and tried to calculate just how much the profits would be. It was clearly a winner, and the figures were in the millions of dollars. During a lull in the conversation Crow turned to Golden and said, "You know, I want to double the interest of your company in these transactions," which meant the local partners would receive a significant windfall. Golden recalled that Crow made the gesture knowing the local partners were even then celebrating their good fortune for what they thought was a good deal.

They had done an amazing job. They were very successful at it, and were happy with what they had. Naturally they were dumbfounded with what they ultimately received. That kind of generosity and support characterized Crow in everything he did, leaving something on the plate for someone else. And what you learn is that it makes such great sense, both in terms of its spiritual element and in terms of its practical element. It's giving, but you receive a whole lot more in terms of how you feel and what you're prepared to do in the future. I think that betting on people and having faith in them has been the key to his success. He has had disappointments, too. We've both experienced partners who have disappointed us in one way or another. Yet he still has that positive outlook that characterizes him.

Almost all Crow's partners, even those who left with some rancor, expressed their admiration for Crow and appreciation for the opportunities he gave them. However, a few departed because they felt their shares were too low, and they wanted to go off on their own to enlarge their operations independently. Then there were a number of partners who retired due to growing apprehensions, either about their own abilities to enlarge upon their businesses or the need to quit while still ahead. Crow was troubled by such departures, but not to the point of considering altering the system significantly. He was keenly aware that sudden wealth could alter perspectives, sometimes negatively, turning hard-working, decent partners into lethargic and avaricious people.

The Stemmonses were Crow's first outside partners, but Gillis Thomas, who arrived in 1959, was the first employee to whom Crow offered a partnership; that is, he was the first working partner, one who hadn't invested money in the company but was given an interest based upon his efforts—what today might be considered "sweat equity."

Thomas was an accounting student at Southern Methodist University when he was hired as a part-time bookkeeper at Doggett Grain. Crow didn't need—and couldn't afford—more extensive accounting services. However, by the time Thomas had graduated Crow was able to offer him full-time employment. Thomas's major responsibility was bookkeeping, but he also did anything else that required attention, which in those days meant assisting Crow with the warehouses. Thomas managed them, began to deal with lenders and real estate brokers, and in fact became a real estate developer. Since Crow still felt he couldn't afford to increase Thomas's salary, he rewarded the young man's efforts by offering him a share of the action in some properties with which he would become involved.

There was nothing formal about the decision. As would be the case with associates joining during the 1960s and early 1970s, Crow simply informed Thomas that he would be a partner, and afterward there was a two-page memorandum to that effect. When Crow expanded into Denver, Thomas took charge of those operations. He remained in Dallas, however, naming others as on-site managers, and in a way he became a training manager, passing on his knowledge to others.

Later Crow came to view Thomas as a hard-working person, but one with limited imagination. For example, Thomas disapproved of Crow's willingness to offer those newcomers partnerships. He would have preferred taking them on, developing them, getting what he could so long as they were pleased with the salary plus commission arrangement, and then, when they wanted more, letting them go. In other words, Thomas wanted to operate like most other companies in the field. Crow rejected such counsel out of hand, saying he wanted *partners* to work under his banner, not employees.

Tom Shutt was the first of the partners to arrive after Thomas had been given a share in some of the warehouses, and his experiences helped establish the trainee paradigm. After graduating from Yale in 1954, Shutt took a summer job in a Dallas retail store owned by a family friend while awaiting induction into the Air Force. Upon his discharge in 1960, Shutt returned to Dallas and a position at the Mercantile Bank. Through a mutual friend he learned that Crow wanted to hire someone to find lessees for the Stemmons Towers. He arranged for an interview, and they hit it off at once. Crow offered him the job at a salary of $1,000 a month plus commissions. Shutt's sole responsibility was at the Towers; he had nothing to do with any other Crow venture. This, too, was to become part of the pattern—partners in one project had nothing to do with the others, and in fact often had no idea of the extent and scope of Crow's total operations. Shutt didn't know what he was getting into, but Crow was equally unclear as to what was developing. He needed someone to perform a specific task. Where that would lead when the project was completed could be dealt with later.

The real purpose of the Stemmons Towers was to create interest in the area around The Trade Mart. The first of the four buildings was going up. With only 3,500 square feet on each of its 12 floors, the initial tower was rather small, so Crow didn't expect it to be particularly profitable, even if fully rented. Crow hated losses on any project, and knowing the slim margin at the Towers, he urged Shutt to work quickly.

Crow didn't attempt to indoctrinate Shutt with his philosophy of management or sales, probably because he lacked a coherent view of both. Nor did he instruct him in how to function in the field, make contacts, work with clients, or anything of that kind. All he did was tell Shutt about the building and indicate the rents he hoped to get. In addition, Shutt was informed that anytime he felt like it he could read anything on Crow's desk to see what he was up to, and that it would be to his benefit to get to know Gil Thomas, who by then had become quite conversant with the warehouse business and understood Crow's way of operating. Thomas had no experience renting office buildings, however, so Shutt was essentially on his own.

Shutt's hiring in January, 1960 was one of those cases in which a person is thrown into the water and told to sink or swim. Shutt swam very nicely, along the way getting into the area of designing offices and practically every other aspect of the job. By mid-summer of 1961, when Crow moved into the Stemmons Towers, Shutt had rented almost all the offices.

Crow thought it fair for Shutt to have a pecuniary interest in the project he had done so much to make a success. One day he walked into Shutt's office and told him, "You're going to be a partner." Along with that change in status Shutt was to move into Crow's office and hire someone else to lease the rest of the Stemmons Towers. The partnership involved the aforementioned 80–20 split. Shutt would perform most of the tasks required on new projects, and Crow would provide the reputation and credibility required to obtain financing. The matter of raising the stake later on wasn't discussed. Neither man knew just where the arrangement would end.

Shutt wasn't sure it would be to his financial benefit to accept the offer; he had no precedents to serve as a guide. The operation really had no track record. The only example of what might come out of a partnership was that of Gil Thomas, who by then was doing quite well with warehouses. Partnership would mean losing out on rental commissions and having to rely completely upon his salary, which wasn't munificent. His rewards would have to come from the marketplace. In transforming from employee into (part) owner, he would have to relinquish the comparative security of the former status for the potential gains of the latter. Even so, after considering the alternatives, Shutt accepted. For the next six months or so Crow really didn't know what Shutt's tasks would be, but Shutt had developed a liking for warehouses through Thomas, and he drifted into that area.

Crow's operation was still comparatively small in the early 1960s. In addition to Gil Thomas and Tom Shutt, there were Hope Hamilton, who had arrived to take care of accounting after Thomas moved on to real estate development, and Barbara Collins, who functioned as a secretary and assumed some of Hamilton's responsibilities when the accounting duties took up all her time. Crow took on additional staff with reluctance, and only after becoming convinced they were absolutely necessary.

Crow hoped that whoever worked in the office would free him to concentrate on deals. He continued to be intolerant of details and constantly sought ways to eliminate or at least limit the way they took up his time. There were all of those meetings with mortgage loan brokers, for example. At first these were crucial, but as Crow developed a reputation they became a matter of rote. It wasn't long before he was asking Hamilton to handle these matters, and in the process she not only became better acquainted with the operations, but emerged as a combination executive assistant and financial officer. Because the business was expanding rapidly, Hamilton's duties and responsibilities increased almost on a weekly basis.

In this period all the ledgers were hand-posted, as was the check register, and the general ledger was posted only once a year for tax purposes. A ledger was established for each Crow operation, whether it was a partnership or corporation, with a separate set of ledgers for those deals Crow owned individually. Engineers, architects, construction firms, and others submitted their bills to Hamilton—in some cases, such as McFadden & Miller, with no breakdown—and she made out the checks, had Crow sign them, and then put them in the mail.

One of Hamilton's tasks was to inform Crow about the status of cash flow and bank balances. She considered it part of her duty to warn him of shortfalls. It did little good. Crow, she recalled, would look pained for a while, slump in his chair and place his head between his hands, but the next day be at it again. Instead of trying to obtain sufficient cash for old deals, he would be entering into new ones for which Hamilton had to try to arrange payments.

Hamilton was intelligent and hard-working, and quickly absorbed the knowledge required for her duties. Even so, there was too much for any one person to do, and Crow kept piling on more and more work. In addition, she had become increasingly concerned with Crow's penchant for expanding without what she considered sufficient capital. Finally, pushed to the point where she no longer could stand the strain, Hamilton told Crow in the spring of 1963 that she intended to resign. Crow realized he would not only have to replace her, but devote some time and energy to administration. The prospect wasn't appealing.

Help was on its way in the person of Bob Glaze. Fair Department Stores, a Fort Worth-based operation, had recently been acquired by Allied Stores. Glaze, Fair's controller and treasurer, disliked the change in management and resigned. He then served for a while as president of Servel Drugs, a chain of five drug stores. Unhappy there too, Glaze was looking for a new job.

At the time, Crow was promoting the Apparel Mart and was seeking agents who could help lease space in it. Glaze learned of Crow's activities through a cousin, Carl Hunt, who lived not far from the Crows. Crow wanted a leasing agent, and Glaze was an accountant with no interest in sales, so it appeared nothing could come of a meeting between them. One was arranged anyway. That afternoon Glaze arrived at Crow's office, and unexpectedly was interviewed by Horace Ainsworth, who was deeply involved in the effort. While Ainsworth attempted to persuade Glaze to become a leasing agent, Crow sat nearby, conducting some negotiations over the telephone and eavesdropping on the conversation.

After a while Glaze left, having decided there was nothing for him there. Two days later, however, Crow called to invite him to lunch. Once again Glaze found himself in an unusual position: Gene Locke was at the luncheon, and Crow spent almost all the time talking with him about several projects.

Afterward Crow and Glaze went to the Hartford Building. Standing in the lobby, Crow told Glaze he would meet him in his office; he intended to walk up the 14 flights of stairs. Nonplussed, Glaze said he would go with him. They talked along the way, and by the time they got to the office, Glaze had the post of vice-president and treasurer at a salary of $16,000 a year. The understanding at the time was that Glaze would be a staff member involved with all the Crow interests—the partnerships in warehouses, the marts, the office buildings—everything. He would have an overview of the entire operation.

It didn't take Glaze long to understand what Hamilton had been up against, and also to realize her value to the organization. He told Crow that he wanted her to remain, and learning of her plans for a European trip, offered to do her job as well as his own until she returned. All three agreed to the arrangement, and with this Glaze began work.

One of his first undertakings was to bring some order out of the jumble of records. When he asked to see the company books, he was presented with that single set of hand-posted ledgers in which entries were made whenever required, with everything toted up once a year. A request for a list of the partnerships brought forth a little black book in which Crow had recorded them, again by hand. Actually Crow didn't have to refer to this record book, since all the information was filed in his head, but Glaze regarded the lack of reliable records as unprofessional, and so he devoted much of his time to formalizing and rationalizing operations.

Glaze's treatment and progression in the company were similar to those afforded Thomas, Shutt, and Hamilton. The more he became acclimated to the work, the more toil and responsibilities he was given. After a few months Crow gave him power of attorney over all of his operations, and Glaze started overseeing them as well. When the Crows decided to take a three-week vacation in Europe, Crow put Glaze in charge of all projects, with the right to buy and sell and enter into binding contracts.

A new pattern was developing. Crow would put together a deal, and when everything was in place he would gradually turn it over to Glaze. As always, Crow liked to work on three or four projects at a time. With Glaze to take care of things he could expand his operations to five or six, knowing that their execution was in capable hands.

Glaze had an uncanny ability to organize the efforts of many people, from accountants to bankers to lawyers, and get a transaction closed and, when necessary, documented. He could cut through peripheral matters and go to the heart of problems and issues. In fact, it was always Glaze's way to arrive at the basics as soon as he could, try to deal with them in a systematic fashion, and then go on to the less-important matters. One of Crow's attorneys, Bob Middleton, who knew both men well, once remarked that Crow was able to operate from the mountaintops because he knew Glaze was coming through the valleys, damming up all the tributaries and getting everything done. What he meant was that in those days Glaze was Crow's alter ego and right-hand man, freeing him for forays outside of Dallas. Several years later Crow would comment that if one could splice Ewell Pope with Bob Glaze, one would have "the key to the ownership of the free world." Glaze remained with Crow for 13 years, and never received a raise from his initial $16,000 a year. But as a partner on quite a number of projects he soon amassed a net worth in the millions of dollars.

As had Hamilton, Glaze regularly pleaded with Crow to establish some kind of organization or at the very least impose a sense of discipline and order in his operations—in effect, to organize a Trammell Crow Company. Crow fought the idea. He had no interest in or time for such matters. All he seemed to care about were those deals, and by then there were so many of them that he had become swamped with work. Had there been 100 or 200 hours in a day, he might have done it all by himself. Under the circumstances, however, Crow had to take on others to assist in the efforts he had initiated.

Glaze was left on his own to manage matters in the office, but Crow refused to consider changes in the way he conducted business. What he really needed was more partners who would not only enter into deals in various locations, but manage the properties as well. So he collected them, not so much because he wanted to do so, but because that was the only way to free himself for negotiating more deals. Additional assistance in the field, then, not structural change at the home base, was his response to the growth of his operations.

Throughout his career Crow preferred to take on individuals inexperienced in real estate rather than seasoned developers. He once said that the perfect partner had intelligence, drive, and knowledge, and that given the first two, he could provide the third. A major goal of his career has been to demonstrate that one can take a person with good native abilities, give him or her the right kind of leadership and inspiration, and enable that person to be productive.

Crow liked to hire young people. He was convinced that youth and energy tend to go together, and that most young people yearn for success. Also, the youthful willingness to meet challenges was usually accompanied by a sportive outlook, a sense of excitement, and an eagerness for new experiences, attitudes that can be useful in real estate development. The people Crow preferred were those who were uncertain enough about their abilities and prospects to work far more assiduously than they had to, demanded success to justify their existence, and felt that they would be failures unless they achieved managerial responsibilities—or in Crow's case, a partnership—before they were in their mid-30s. By the time they were 40 or so, Crow maintained, some achievers tend to become conservative and averse to risk, or simply desire different kinds of challenges. Then it might be time for them to go off on their own.

One who epitomized this young partner profile was Bart Brown, a Yale graduate who started out leasing warehouses in 1966. From there he progressed to handling Crow's commercial interests in the New Jersey region. Brown would go on to a partnership, and within less than a decade became one of the most significant figures in the organization. He retired in 1980 to pursue private interests. His was a career in which all benefitted.

If everything worked out well, Crow would have a cadre of senior partners in their late 30s and 40s, and many more juniors in their late 20s and early 30s. As the seniors moved on, the juniors would rise to take their places. It was not in anyone's interest for most of the seniors to remain into their 50s and 60s, since this would block advancement for the others. Crow believed that most equally competent older people lack the enthusiastic drive required in the business.

Crow expected loyalty and decency on both sides of the partnership, which would continue so long as both partners benefitted from the arrangement. But the loyalty is provisional, contingent upon the circumstances of the relationship. Crow has been known to dismiss young, ambitious people who have not worked out and to implore veterans intending to retire to stay on a few more years, if only in a diminished role. Thus, while friendships lasted for a lifetime, partnerships were limited by time, ambition, and mutual interest.

Later on Crow would learn that cutting off relationships was not that difficult and need not be tragic. There are few corporate decisions more foolish than to hold on to relationships and operations out of sentiment, sloth, or indecision. The general rule, Crow thought, might be that if you would reject the opportunity to initiate an existing relationship today, you might consider whether it would be wise to end it tomorrow. If the answer prompts separation, the dissolution should be done in a way that preserves the integrity and respect of all concerned.

Crow also believed that impecunious individuals work harder than those who are relatively well off. The reasoning is plausible enough: Ambitious people generally work harder before they achieve a certain level of financial success than afterward, though there are quite a few exceptions in the Crow organization. Paradoxically when those young and hungry associates become middle-aged and financially comfortable, their performance tends to suffer. Attention turns to matters of vacation, leisure, and the preservation of accumulated assets rather than aggressive expansion. It is then they should either retire, go off on their own, or, later on, when operations became more structured, be moved into a quasi-managerial position where they can function as mentors and critics, engaging in strategic planning where their experiences could be useful, rather than getting business and creating deals.

Crow thought he saw the trend to conservatism in Gil Thomas, who after a while became more interested in protecting what he had than going on to meet new challenges. It was the rather common circumstance of a person wanting to leave the game while on a winning streak. Success can make a person bold, but a string of victories may also foster conservatism.

Thomas had demonstrated little of the sense of excitement or the other qualities Crow prized when he first came to Doggett Grain, and by the mid-1960s what drive he had had vanished completely. The fact that he was good at details and that their association was a long one blinded Crow to Thomas's shortcomings, and so he permitted the relationship to drift. It was one of his managerial flaws in this period.

During the early 1960s Crow's interests in the Dallas residential business intensified. The city was booming and naturally was periodically short of affordable housing. Before and immediately after World War II most residential building centered around apartment houses and detached, single-family dwellings. Then, as a result of Federal Housing Authority guidelines, garden apartment complexes became more popular. Zoning concepts changed as well, while financial organizations looked more favorably upon multiple-dwelling operations. Soon large-scale developments made their appearances.

It seemed like a good time to plunge into residential development. As was his wont, Crow looked around for likely associates. His search brought him into contact with Bill Jones, an architect, and Willard Baker, a lawyer, both of whom wanted to enter the business. Together they organized Baker, Jones & Crow, which started out by building apartments in Dallas and then entered additional southeastern markets—Oklahoma City, Austin, San Antonio, Fort Worth, Waco, Shreveport, Tyler, Longview, and others. The focus remained on Dallas, however, which as a result of the boom sparked by the opening of the Stemmons Expressway and other transportation improvements was in the midst of its greatest period of expansion. Jones dropped out after a short time, and the company was renamed Baker-Crow. By the early 1970s it was one of the most rapidly growing residential companies in the nation.

Crow also entered the residential field in association with John Eulich, a dress company salesman who had dabbled in real estate. Crow met him when working on the Apparel Mart. They operated together profitably until 1963. This relationship was not successful. Crow claimed Eulich was not faithful to their partnership and forced a dissolution. Winthrop Rockefeller, with whom Crow had been involved in several profitable projects, bought out Eulich's interest. Eulich later organized Vantage Corporation, which became one of the nation's largest real estate developers.

In 1963 Crow was looking for someone to run the residential business, and he found him in Mack Pogue, a young, aggressive Dallas real estate broker. Pogue knocked on Crow's door one day and said he wanted to sell him some land at Spring Valley Road. He and Crow met occasionally thereafter, but they didn't transact business until two years later, when the first stage of the Apparel Mart had opened and Crow was deeply involved not only in other Dallas projects but also large-scale deals in other parts of the country. It was then he remembered Pogue, and asked him to come to his office to discuss some ideas he had been toying with. Crow wanted to know if Pogue would be interested in joining his residential real estate operation. Crow would provide the financial contacts to raise most of the required capital and would throw in the properties he and Rockefeller owned together, while Pogue would be in charge of day-to-day business.

Pogue was struck dumb by the idea. He was, after all, a real estate salesman, with no experience in development. He took a week or so to consider the offer, and when they met again it was obvious Pogue had given the idea serious thought. He didn't want to join the Crow organization, such as it was, preferring instead to work with Crow in a separate company. This suited Crow, who by then had much experience in such ventures. Since it was close to Lincoln's Birthday, they decided to call the company Lincoln Property, and for the next 13 years they built apartments across the country.

Dallas had an oversupply of housing units at the time Crow and Pogue formed Lincoln, but these had been absorbed when they were ready to rent. Other builders had left the business in dismay, so for a period of three or four years they had the field nearly to themselves. Yet each opportunity also presented problems.

Lincoln started out small, with units in New Orleans and then Dallas, just to get the experience. Then Pogue and Crow went for something more ambitious, a garden apartment/retail shopping complex in Dallas. Pogue not only had never built anything like this, but he hadn't ever managed properties. In fact, neither man had any experience in what they were doing. They did know real estate, however, especially that key ingredient: picking the right location. The site they selected was Willow Creek, not much of a place then, but today a prime residential area due to what they did there.

The land cost $2 million, all borrowed from First National Bank. Pogue hired an architect, Dub Miller, who designed a 350-unit garden apartment complex together with small retail shops. They next went to Connecticut General to obtain the money needed to erect the structures, which were soon completed.

Willow Creek was not a great financial success. Lincoln had made the apartments too large and put in too much quality for the intended prices. When Pogue asked higher rents to cover the costs, potential renters stayed away in droves. So he lowered rents and hired an agent to take charge of the property. Lincoln ate losses for a while, suffering through a negative cash flow for about four years. Crow wasn't particularly discouraged, however. He always believed that if you can hold on, property values will rise and make you look like a genius. That is what happened at Willow Creek.

One of Lincoln's most ambitious undertakings was "The Village" in Dallas, started in 1969 on Southwestern Boulevard, between Greenville and Skillman Avenues. Comprising more than 8,000 units on 300 acres in North Dallas, The Village was one of the most ambitious attempts at suburban planning in this century. There were clusters of apartments for several categories of renters—singles, newlyweds, parents with children, and older couples whose children had gone off on their own. In the complex were a golf course, a shopping center, playgrounds, a club, and other amenities.

Lincoln continued buying land and building; within two years had 17 projects under way in Dallas. For a while the two men concentrated on the market there, the one both understood well. Yet even then they realized it made little sense to focus all their attention on Dallas, when there were so many opportunities elsewhere. The trouble was that neither of them knew anything about residential markets in other cities. It was one thing to go in with Frank Carter and John Portman in a large-scale commercial venture combined with hotels—they and others in the deals knew Atlanta, Crow knew financing, and between them they understood the way commercial renters operated. This personal knowledge base was not available for the expansion into new residential markets. Yet the promise was so enticing that they overcame doubts and went ahead.

In 1968 Crow invited Pogue to his office and told him that while Pogue's initial position as Crow's Dallas partner in residential real estate development seemed to be mutually satisfactory it made sense to expand Lincoln into another city. Houston might be the next move. Pogue liked the idea, and they built a garden apartment complex in Houston the following year, followed by projects in St. Louis and then San Francisco. Each city posed different difficulties, and Pogue had to keep on top of things by flying in and out of all of them on a regular basis.

They soon concluded that conducting business by airplane was untenable; some structural change was needed. So Pogue took three of Lincoln's top Dallas partners and relocated them in Houston, St. Louis, and San Francisco. Each man was now a regional partner, with satellite companies under his control, and a large degree of autonomy.

Crow originated the notion of having regional partners from the Lincoln Property experience and projects with Shutt and others. It was an answer to the possibilities created by success and the enlargement of enterprises. He would approach one of the bright young people who had become a partner outside of Dallas and ask the person to become a regional partner, meaning that he would be in charge of operations in a specific geographic area. The regional partners had a great degree of latitude in their actions, reflecting Crow's belief in permitting associates—partners, that is—as much freedom and leeway as they thought they could handle. Each regional partner (not vice-president) operated largely independently. They kept in touch through telephone calls and occasional visits. Otherwise they were on their own, just as Pogue was with Crow at Lincoln Property in Dallas.

One of the reasons Lincoln progressed as it did was the relationship that evolved between Crow and Pogue. Virtually all men and women in executive and professional positions feel the urge, at one time or another, to act as mentors, while younger people appreciate the help and guidance provided by older hands. Crow was not prepared to step down and play that kind of role in his commercial activities, but he willingly did so at Lincoln, which Pogue ran on a day-to-day basis. For a while, at least, Crow and Pogue had that counselor-protégé association, though it quickly ripened into something else—friendship between equals. They would frequently drop in on one another just to talk. When Pogue or Crow took vacations, the other might swoop in for a day. They would jog along a beach, loll on the sand, and talk about business.

In *Dallas Architecture, 1936–1986,* David Dillon, who believed that "the Trammell Crow Company has become the most successful real estate developer in the history of the earth," paid tribute to Crow's development of people as well as buildings. Dillon observed, "Through its [Trammell Crow Company] doors have passed several generations of Dallas developers, including giants like Mack Pogue of Lincoln Property and John Eulich of Vantage Corporation. A major chunk of Dallas has been built by individuals who went to school at Trammell Crow."

In the early days most new employees were those whom Crow met in the course of business. He had a chance to observe them in action, talk with them informally, and so could make a judgment before making an offer. Mack Pogue was recruited in this way, as were many others.

Crow wanted to be surrounded by associates who had original concepts of how, where, and why business should be conducted. He respected those who struggled to arrive at their own conclusions, even when the beliefs differed sharply from his. Crow sought and prized originality and enthusiasm more than conformity and agreement. For a person clearly the senior figure in all his operations, he was always surprisingly ready to consider concepts that differed from his own, especially when they originated from young trainees.

Those early partners came from a wide variety of backgrounds, but there were some common denominators besides youth and ambition. One was the matter of personal habits. Crow didn't smoke and rarely drank, and he preferred associates who were the same in their behavior, though there were a few smokers and social drinkers in the organization. Even so, a potential trainee had to demonstrate some concern with keeping physically fit and a regard for the feelings of others.

A sense of playfulness was also a plus. When delivering a talk before a class of M.B.A. candidates, Crow was asked by one student his opinion of designer jeans. He gave the matter some thought, and replied, "First of all, if a designer jean company wishes to rent any warehouse or exhibition space from us, we'll be happy to oblige them."

"Second, I don't wear them myself."

Then he leaned forward from the podium and affixed the questioner with a look of mock severity.

"And third, nobody who works for me wears them!"

Obviously there were more significant partner characteristics than fashion preferences. Like all businesspeople, Crow wanted "the best people." What did he mean by that? The key wasn't academic performance, though that was taken into consideration. High ethical standards were a priority matter. While Crow believed everyone in his company worked and sold and pushed as hard as possible on every deal or project, pervading his thoughts was the desire to take on people who would not compromise his ethical beliefs, especially the tenets of trusting people and treating the customers well. And in keeping with his view of work as fun, he wanted to recruit people whom he and his associates enjoyed being with. To be sure, an enjoyable personality is an intangible factor, but one that was valued due to experience and judgment.

Looking back at the experiences of accumulating partners over three decades, Crow attempted to describe just what it was he had been seeking.

First off, I never hired anybody who wasn't a good guy. A fellow, I used to say, someone you'd like to have a beer with. That's number one. Then I also wanted brains. I would try to assess the person's intelligence. The next thing I would say [is] does this man perform as he should when he has to or will he not do so. Is he self-disciplined? Those are the most important factors.

I would sit there and make a judgment. For a while I gave psychological tests to people, but that was soon abandoned. There is no way to test for these kinds of qualities. Of course I made some errors, but at least they were my errors, and not those of someone who didn't understand what I wanted and needed in a partner. I expect that for every one of those people who's with us today we probably had three or four who came in and didn't make the cut. They either left on their own or we let them go.

Margaret Crow summed it up this way: "Trammell seeks out people with high personal standards. He picks disciplined people. He has always tried to work with bright, nice people. Trammell thinks nice people are more likely to be successful people."

Locating "nice people" wasn't a matter of happenstance, though Crow strove to cultivate that impression. In reality, most of what he did resulted from careful calculation. Take that trip to Harvard at the invitation of Howard Stevenson. For the first time in his life, Crow found himself surrounded by dozens of intelligent, sophisticated, aggressive young people eager for business careers. What a wonderful place to recruit salespeople and future partners! Not only were they superbly prepared for the kind of training Crow had in mind, but through Stevenson they had learned of the fortunes awaiting successful real estate developers, and that these derived from the market, not salaries. Crow might take on a few of them at low wages, holding out the promise of a partnership and great wealth if things worked out to their mutual satisfaction.

In those years Crow started newcomers at a salary of $12,000, which was low by the standards of the time but sufficient in his opinion to lure them to Dallas and other offices around the country. At first it was difficult to get recruits at such a price. Not for long; soon the graduates lined up for jobs at Crow, knowing how well their predecessors had done. This was still true in the late 1980s, when Wall Street was paying over $100,000 a year for some selected Harvard M.B.A.s, while Crow offered $32,000 plus commissions—and dreams. So after a while Crow had his choice of the lot, and could be selective, setting down criteria he and his associates deemed significant. Of course not all those around Crow were partners, or even wanted to be. Crow appreciated the importance of diversity, and so attempted to acquire those who could be useful in a variety of ways.

Not only was the work difficult, requiring long and irregular hours, but the newcomers also had to adjust to a kind of atmosphere many must have found alien. Crow's partners had to be capable of functioning with a highly unpredictable and often idiosyncratic man. They had to adjust to and accommodate his methods of conducting business, which posed problems for some of them. It was not the kind of place many young people would function in effectively, which is why it was so critically important to identify and take on the proper trainees. One way to assist in screening recruits would have been to let the world know about the criteria and the conditions, so that unsuitable individuals would simply look elsewhere for positions. Crow rejected the notion, however, preferring to find his people fortuitously.

By the late 1960s Trammell Crow was an important regional power in real estate development who had made his impact on other, selected markets. Yet little was known about the man and his operations. Except for the Dallas press, he was not mentioned frequently in newspaper and magazine articles. This lack of publicity was the way he wanted it. Crow held that at times it suited his work to be viewed as a sort of "mystery man." It was a stance he adopted early and maintained later on because he felt it provided him with an aura that piqued curiosity and helped business. It is a calculated policy, as he observed years later.

For years I shunned publicity, agreeing to magazine articles with some reluctance, and not even having a public relations representative until quite recently. The reason is—or to be more precise, was—that I believe businessmen would do well to allow some mystery to surround their activities. This is something I learned quite early in the game. It often pays to cultivate an aura of enigma, to swoop down into a meeting in unexpected ways and put people off guard.

This is not to suggest that Crow behaved as he did as a matter of artifice. But he recognized the worth of his natural inclinations. He added, "I have been accused of many things, but never of being a bore or a phony. If you are interesting and decent, don't think there is a need to dissemble. In fact, the attempt can be self-deceiving."

Often Crow seemed to go out of his way to be perverse. One of his beliefs was that businesspeople shouldn't attempt to explain everything they do and need not seek to justify their every action. On occasion they should make decisions without consulting others, and then throw out instructions for their implementation as though over their shoulders in a half-serious fashion. Again, this offhandedness was a natural part of his personality.

Not all decisions have to be shared with all associates, Crow believed. This is not to say that he was secretive; as has been seen, he invited associates to inspect papers left on his desk. But Crow thought it best to keep his own counsel and mete out information to others on a "need to know" basis. There are times when one should tell others more than they might want to hear, and on other occasions they are told less than they wish to know. He was not consistent, even when it came to dealings with partners. This made Crow a difficult person to work with, as his associates often remark.

What all of this came down to was that intuition, informed by intelligence, and calculation ever were Crow's hallmarks. Outsiders tended to see the first of these qualities, and not to be fully aware of the other two. That was the way he wanted it. And what Crow wanted, he usually was able to get.

His distaste for structure, for example, meant that so long as he was in charge, there would be as little of it as possible. The expansion continued without any formal organization in place. Crow extended operations into Houston, having purchased some land from the Southern Pacific on which he had erected a number of warehouses. For a brief time he worked through John Eulich, and when Eulich left, Shutt was asked to take over in Dallas. Crow hired Bobby Wilson to go to Houston and be his representative on the site, taking over the properties there after the split with Eulich. At the time Crow had around 500,000 square feet of warehouse space in Houston. Shutt and Wilson put up an additional 200,000 to 300,000 square feet a year for the rest of the decade.

Other expansions occurred in an equally serendipitous fashion. For example, Crow decided that the St. Louis market was ripe for warehouses, and, knowing they would need the services of someone on the scene who understood the market, Shutt contacted a local real estate broker, Bob Kresko, whom he had met through his brother several years earlier. Shutt telephoned Kresko, told him the Crow interests were coming to St. Louis to build warehouses, and asked him to prepare a list of promising properties. This was the first step in an arrangement that, by 1967, accounted for 75 percent of Kresko's business. Almost all the buildings subsequently erected were warehouses, usually 20,000 to 70,000 square feet, leasable in units of 5,000 square feet.

More than a year was to pass before Kresko met Crow; all the dealings were through Shutt. Kresko became a partner in 1967, but even then he met with Crow only once or twice a year. The relationship went smoothly. Kresko was performing well, so it wasn't surprising for him to be asked to help develop other cities as well, starting with Memphis, and then on to Cincinnati and Kansas City. By the early 1970s Kresko was finding it increasingly difficult to handle the entire territory, so he worked in Memphis through a local broker, Tom Farnsworth, who in time became a Crow partner there. Other, similar arrangements followed.

So Crow had recruited Shutt, who recruited Kresko, who did the same with Farnsworth. In this way the Crow commercial business was developing into something that at first glance seemed to resemble a pyramid, but it would be incorrect to consider that the organization had so formal a design. Rather, it remained a web, with Crow at the center, and growth was shaped more by intuition than anything else.

Thus in an apparently random fashion Crow and his partners had established an organization of sorts, and added a third tier to the compensation hierarchy. In a typical venture, one of the partners would work out a deal in conjunction with the home office. Crow would get his percentage, then would come Shutt's, and the rest would go to the other partner. By the end of the 1960s Crow had what amounted to a group of regional partners.

There was something feudal in these arrangements. Crow was ensconced in Dallas, conducting his varied interests, and in the field were the barons, each in his own territory. They were erecting baronies of warehouses, hotels, residences, office buildings—whatever might be promising for their areas. The partners used the Crow name and credit, but did so with local financing institutions. All were semi-independent, in that they had a great deal of leeway in their actions but they had pledged their fealty to Crow. All the contracts signed under the Crow name had provisions for financial rewards to Dallas.

Thomas and Shutt set part of the trainee development pattern that prevailed until the mid-1980s. Trainees started out with small assignments and expanded as rapidly as their talents and interests permitted. Crow tried to give these associates as much freedom and leeway as they felt they could handle. During the first few weeks at the company, a newcomer tagged along with one of the older hands for orientation. Then he was set free. The senior people were there for guidance, information, and assistance. For most of the newer partners, this assistance usually amounted to very little.

In those days (and later) a new associate started out as a leasing agent, for Crow believed this the best training ground. This was strikingly different from the paradigm employed at many other development firms where newcomers start out in construction, finance, architecture, and other related areas. Crow associates and partners know even before arriving at their first assignment that all deals begin with the customer, that everything flows from the tenant's requirements.

Initially leasing agents made "cold calls" door-to-door, dropping in without prior knowledge of the prospect's needs, rummaging for news of possible moves within or into the agent's area. Cold calling was a rude shock for quite a few of the M.B.A.s from prestigious universities, but Crow saw it as a salutary experience. After seasoning, associates were given leads, and in time developed their own clientele. The leasing business, then as now, was a one-on-one process, and a person who cannot handle the fundamental activity was unlikely to succeed in other aspects of the development business, especially in training and tracking others when the time came to do so.

Every move a developer makes in operations such as warehouses or marts finds its success or failure in its acceptance by tenants, said Crow. Lessors really couldn't perform surveys to learn customer preference, or test market a product or area. But the decisions on what land to buy, what to build, the architectural dimensions, the style, the level of cost, and a hundred other matters had to be made with current knowledge and informed guesses about what client-tenants would accept. Every decision maker in all stages of development was fundamentally a leasing agent.

Crow associate Tom Simmons didn't understand the importance of learning the leasing business when he arrived in 1970. After graduating from Harvard Business School, he had gone to work for Ernst & Ernst in Washington, D.C., but he was soon dissatisfied with that position. A North Carolinian, Simmons wanted to live in the South, and after casting about he decided Atlanta would be a good place to seek employment. He telephoned a friend there and obtained a few leads, one of whom was Frank Carter. While there were no openings then at Crow, Pope & Carter, Simmons was told, there might be something available with John Portman. Portman likewise had no need for a trainee, but he sent Simmons's letter on to Bob Glaze. Several months passed, and Simmons was becoming increasingly unhappy with both his circumstances and prospects. Then came a telephone call from Glaze, inviting him to Dallas for an interview. Simmons was not impressed with what he was told about the operations, but his wife's grandmother lived in Fort Worth, and he had never been to Texas, so he decided to make the trip. He was interviewed by Glaze and met Crow, who decided then and there he wanted Simmons to come in as Crow's administrative assistant and Glaze's financial assistant. Recalling the interview, Simmons remarked:

Trammell's kind of funny. He assumes that if you are interviewing with him that you are there because you want to go to work for him. So if he offers you a job, he simply assumes you are going to take it. I remember he kind of pressed me for an answer, but I did want to talk it over with my wife. Bob Glaze understood this, and told Crow, "Let's give him a day or two to get back to Washington and then get back to us." But I knew I was going to take it.

As it happened both Crow and Glaze were traveling quite a bit in this period, so for the first month Simmons had little to do but wander around the office and try to learn all he could about operations, wondering when he would start working with Crow and Glaze.

Then Crow swept into town and embarked on the project to construct 2001 Bryan Tower, Dallas's first skyscraper erected by a single developer, an efficiently constructed edifice that was to set the style for others that followed. In his office, Crow spotted Simmons, who was the only person unattached and available. He would help in the creation of the major new office building. The fact that Simmons had absolutely no experience in such work didn't faze Crow. Simmons became an important player in this project. He recalls that in introducing him to the architects, bankers, and others, Crow remarked facetiously, as Simmons was an athlete and quite large, "I'm going to do the heavyweight fighting and Simmons here is going to do the lightweight fighting."

After a while Simmons was out in the field, leasing warehouses. When an opening appeared in Houston he quickly took it, and went on to assemble operations in the northern and eastern parts of the country.

The training—or lack of it—and the tasks assigned seemed to be matters of chance. Perhaps so, but a few years later, when Simmons was traveling with Crow in England, the two discussed the older man's methods. "He said that really one of his major goals in his career has been to prove to people that you can take a young man who has good abilities and with the right kind of leadership and inspiration, could make him productive and profitable."

As time went on Crow added partners as though stringing a necklace, and almost always in an apparently impromptu or extemporaneous fashion. As Glaze's experience indicated, whenever the opportunity presented itself, Crow opted for an unusual interview, probably to discover how the candidate reacted to unexpected circumstances, in the hope of gaining some insight into the applicant's character. For example, Gary Shafer arrived in Dallas for his interview in 1966 broke, with a recent M.B.A. from the University of Wisconsin and an appointment to see Bob Glaze about a job. Crow, who has always disdained private offices, eavesdropped as usual. Shafer has no difficulty recalling the experience.

He had one foot on the desk, and one foot on a little chair he kept next to him, and he had a telephone at each ear. He was being real nice to one guy and he was giving the other a hard time. Trammell would put his hand over one mouthpiece and talk pleasantly, and then he would put his hand over the other and blaze away. All the time I was trying to keep focused on Glaze's questioning. Eventually Trammell got off the phone, and he stood up and looked at me from under those bushy eyebrows. He also had a band-aid taped right down the middle of his forehead. He looked at me and he said, "What on earth are you doing here?"

Neither man can recall the answer, but it must have been what Crow hoped to hear, because Shafer was offered a position on the spot.

He did not accept the post at that moment, however. Certain that he wanted to be involved with real estate development, Shafer already had an offer from Rouse, another developer smaller but better known than Crow. He had to choose between the two of them. In considering the matter, Shafer concluded he would be more comfortable at Crow.

Well, they're quite a bit different. Rouse Company was more formalized, it was more organized, and gave the appearance of being a lot more polished. It probably was more professional if you were to use those terms. The Crow Company at that time was mainly just Trammell and a few other guys, it wasn't that big a deal, and came across as more casual, but not laid back. They were every bit as aggressive if not more so but less formal.

Shafer's initial experiences were not unlike what Simmons went through. Within a short time he was busily involved with several projects in which he was more or less given a free hand. "In many cases I almost had to force myself upon Bob or Trammell to get their input in it. It was complete laissez-faire." Shafer's partnership arrived equally informally. "It just kind of happened. There wasn't any pomp or ceremony or drill to it. There wasn't any covenant, deal, or anything else. It was just kind of here are the papers. At that time it was a little two-page partnership document and off we went."

Don Childress came for an interview in 1971, fresh out of the University of Texas Business School. They talked for a while, and then Crow said, "I think I want to hire you and I want you to work in Dallas." Childress wanted to know if that meant he had the job. He did, but Crow desired to learn how badly he wanted to come in, so Crow told him, "At your age, you don't need to work. Don't do anything for a week." Crow telephoned Childress at 6:30 A.M. exactly a week later. "Are you ready?" he asked, and Childress said he was. "Be here at 7:30." He arrived on time, and went to work that day.

Take chances and learn to trust your informed hunches—just as this was Crow's way in entering into deals, so it also applied to taking on newcomers. With deals, the conventional wisdom is to study a situation from every angle before coming to a conclusion. Crow's associates sometimes acted on instincts and were rewarded for it. In the late-1970s Childress, who by then was located in the Atlanta office, had the opportunity to acquire a choice site at a reasonable price. At the time he wasn't certain it could be marketed, or even what kind of project it could support. Childress realized that it was a unique parcel, however, the kind that didn't come along every day. So together with regional partner Tom Simmons he made the commitment, using almost everything liquid in the Atlanta office as a down payment. Then he searched for a joint investor to share the venture. He didn't locate one, and was obliged to ask for an extension of the deadline, placing even more money at risk. Now not only was the office at stake, but his own future was on the line. Childress finally found his investor, one willing to sink $12 million into the purchase, which ultimately became the famous Galleria.

Tom Bailey arrived in 1971 in an unusual way. In recalling the event, Bailey makes it seem to have been almost against his will. A recent University of Texas Business School graduate working in Los Angeles for an Austin venture capital firm, he received a telephone call from a former professor asking for a strange favor. Crow was coming in to conduct a few interviews on brief notice, and the professor was short of candidates. Would Bailey impersonate a job seeker?

Bailey agreed, but he had to attend a Marine Corps Reserve meeting first, so he arrived at Crow's hotel wearing his khakis. Crow sensed Bailey had what it took to make it as a trainee, and soon offered him a job leasing warehouse space in Dallas. Whereupon Bailey somewhat shamefacedly confessed the imposture, assuming this would end Crow's interest. The deception was so wild and innocent Crow couldn't help laughing, and being fascinated. If Bailey could do such a good job of selling himself when he didn't want a job, how would he perform when he really wanted something? So Crow repeated the offer, which was politely rejected.

Bailey's reluctance only whetted Crow's interest. When he returned to Dallas Crow telephoned the young man to try to convince him to visit Crow there. He wouldn't accept, but Crow didn't intend to give up. Finally Crow asked him, "Tom, why aren't you beating on the door to work for my company?" This must have been the right approach; by the end of the day Bailey agreed to accept a position in development under Gary Shafer in Houston.

The hiring of Hayden Eaves, who came in from Los Angeles in 1973, was more typical. Crow and Eaves talked about the younger man's family, children, personal values, morals, and his aspirations. Then Eaves was taken to some of the other partners, returned to Crow in the afternoon, and was told, "I would like you to be part of my company." That was all there was to it.

In those days each partner was involved only with his own operations. Communication among partners varied from poor to occasional to nonexistent, which led to embarrassing situations. For example, when Bailey first reported to Houston he went to the Crow offices there, expecting to be informed about what was required. Not only was there no place from which Bailey could work, but the staff hadn't been expecting him. In fact, they had never heard of Bailey. Crow had forgotten to tell them about the new man in town. This lapse did not mean there was chaos in the field; in fact, commercial operations were proceeding very well. But there was a lack of coordination and of an established means of disseminating information. In this instance Bailey operated on his own.

Crow expected the young people he took on to be able to work things out for themselves, as did Bailey. To illustrate this, consider the experience of Sandy Gottesman, who in the early 1970s was a newly recruited associate in the Austin office. Crow had 1.5 million square feet of available space in Austin then, and local business and government agencies had leased only about 300,000 square feet. It was a difficult period for Gottesman, and no one volunteered advice or assistance.

Crow owned a 160,000-square-foot facility in the Austin suburb known as Exchange Park. Activity there was so slow that the empty parking lot embarrassed him by advertising his distress. At one point Gottesman considered inviting Hertz to use the lot free of charge just to make it appear the Park had tenants. He decided instead that his best chance to rent the facility lay in persuading Electronic Data Systems (E.D.S.) to house its medical claims processing operations there, providing E.D.S. won the state Medicaid contract then out for bid.

Gottesman monitored the bidding, and when E.D.S. was awarded the bid, he pursued his objective. He even went so far as to stake himself out in a phone booth next to the auto rental company that E.D.S. President Mort Meyerson patronized on his trips to Austin. Then he learned that Chairman Ross Perot would be coming in to make the final decision on location. Gottesman got a close haircut, took out his wing-tipped shoes and white button-down shirt, and prepared to meet with Perot. He obtained a 2:00 P.M. appointment with the E.D.S. chairman, but was stood up. After waiting two hours, he called the local E.D.S. office and learned that Perot had been detained and was at that moment on his way to the Austin airport to catch the 4:30 P.M. flight back to Dallas.

Gottesman got into his car and rushed to the airport, where he reserved the seat next to Perot. For almost an hour he had Perot's total attention. Gottesman won the contract, and soon after he was promoted and became Crow's new partner in Austin.

Many of Crow's people exhibit that kind of originality. Another leasing agent, Steve Hanna, was frustrated in getting an appointment with a potential client. Telephone calls, letters—all were ignored. So Hanna composed a singing telegram and sent it to him. That strategy worked.

There are certain things Crow does not do when seeking associates. He has never brought in anyone on the recommendation of a "headhunter," and he does not employ psychological testing. He concedes such methods could be appropriate for other organizations. Perhaps the people he took on would have performed superbly on the tests and would have been recommended strongly by headhunters. Perhaps not. But Crow couldn't afford to take the risks involved in not trusting the judgments of those who know the most about his needs—himself and his top associates.

Crow had experimented with testing devices before discarding them as alien to his cause. In 1973 he tried using psychological examinations. One young man, Joel Peterson, was interviewed by him and some others, and then escorted into a room to take a battery of tests designed to determine his suitability for the development business. The tests contained questions Crow thought were rather personal. This bothered him, for his instincts told him that this six- to eight-hour ordeal really wasn't necessary, since he had already decided Peterson would be a fine addition to the firm. By then Crow was becoming involved in Europe, and could use a person with the interviewee's background. So halfway through the tests Crow walked into the room and told Peterson, "Forget about this. You've got the job. Don't worry about these tests. I want you to become my partner on the French Riviera." His salary would be $14,000, not very much even for 1973, and as it turned out it was the lowest Peterson had been offered by anyone at the time.

The decision to hire was based on more than simple instinct. Peterson was well equipped for the tasks. He had spent two years in France as part of his Mormon missionary obligation, and in addition to being fluent in French, he had a keen understanding of the economic system and business methods there. At the time Crow had big plans for expansion into France, so Peterson seemed a fine acquisition.

Peterson knew more about the Crow operation than most newcomers. When Crow visited the Harvard Business School and told the students that the most important ingredient in business was love, Peterson was one of the students in the audience. A friend of his, Randy Paul, had taken a job with Crow the previous year and was assigned to Austin. Paul (who later became Crow's partner in Salt Lake City and then Phoenix) tried to interest Peterson in real estate, but at that time he was more concerned with a career in international business. Perhaps the mention of the Riviera made the difference; at any rate, Peterson accepted Crow's offer.

Crow intended Peterson to spend six weeks or so in Dallas to become oriented to the business before leaving for France. Those weeks became months because there were some difficulties in the French start-up. Finally he flew into France. Peterson, his wife, and new baby lived in Paris for about a year on a small salary plus commissions, working on two large office blocks. Then they went off to Lyons to handle a warehouse project. After eight months, he was supposed to become Crow's regional representative, working under the direction of the local partner, Gil Rogier, located on the Riviera. But Peterson never got there.

Peterson was not very happy in France. He felt trapped, thinking it a dead-end job. He was displeased with his colleagues there, feeling they were not particularly loyal to the firm and weren't doing the kind of work required. Peterson was young, aggressive, intelligent, and ambitious. He saw no future for himself with this aspect of Trammell Crow. In addition, he was receiving job offers from other firms and was seriously considering a move. Several special assignments followed, however, that made him decide to stay on.

In Peterson's case, Crow had needed a person for his French operations and a qualified applicant happened along. Had Peterson arrived a few months earlier or later he would have had a job, but probably at something quite different. Often, when Crow had a slot to fill, the next person who came along and seemed to have the necessary personal and professional qualifications got it. Such was the case with Allan Hamilton, a former Navy flier. While in the service he had taken a few courses in real estate at night school and through correspondence programs. Upon discharge in 1964, Hamilton looked for work, and since his wife wanted to remain with her recently widowed mother in Corpus Christi, he aimed for a post in Texas.

Hamilton finally arrived in Dallas, and managed to get an interview with Tom Shutt who introduced him to Bob Glaze. He and Glaze were talking when Crow walked by. Crow had just completed the takeover of George A. Fuller, and was thinking of the need to staff the company. At that moment Crow heard Glaze say, "Mr. Crow, here is a young man who is just getting out of the Navy and wants to go to work in the real estate industry." They shook hands, and after looking the flier over for a moment Crow said, "You want to be a real estate man, do you? You know, it's not that easy." Hamilton replied, "Yes, sir. I realize that. But I've been thinking about it for a long time and I've taken a number of courses. It's not really something I thought about yesterday." Crow asked him about his schooling and grades, and was told they were a little above average.

"Do you smoke?"

"No, sir."

"Do you drink?"

"Yes, sir."

"Do you go to church?"

"Not very often."

In between questions Crow telephoned his secretary and made reservations to go to Brussels. He was certain Hamilton thought he wasn't paying attention to him, but in fact he noted his every reaction. Crow fumbled around, signed a few checks, and could sense Hamilton was getting annoyed but handling himself well.

"How much money are you making in the Navy?"

"Well, a little over $11,000 with flight pay."

Crow feigned shock. "Oh my God. We can't begin to pay you that much. Are you married?"

"Yes sir, with a little boy."

"We can start you at $600 a month. I think we've got a position for you. First thing in the morning I'd like you to go down and visit Cloyce Box. Cloyce is going to be chairman of the largest construction company in America and he needs somebody to be an assistant and fly around in the airplane with him. And we'll have some real estate things for you to do, too."

Hamilton did well at Fuller even though that company proved a failure for Crow, and he gradually edged his way into real estate, handling some of Crow's interests in Chicago in much the same way as Kresko functioned in St. Louis. And, just as Kresko expanded from his base into Memphis and Cincinnati, so Hamilton entered Milwaukee and Minneapolis. In time he would become a partner and the developer of one of Crow's major ventures, Hamilton Lakes.

In 1969 Lincoln Properties opened an office in San Francisco, and Preston Butcher, whom Pogue had recently taken on as a developer, was looking for someone to represent the firm in that market. He soon located Ned Spieker, who seemed an ideal candidate. At the time Spieker, a University of California at Berkeley graduate, was engaged in development work for Dillingham Corp., the large Hawaiian conglomerate, and was interested in a move. But not to Lincoln; Spieker was more concerned with commercial properties. Butcher persisted, and asked Crow, who was traveling in the area, to see if he could assure Spieker of the wisdom of the move. The more Crow talked, the more Spieker became convinced he should be working for him, and not Lincoln. Crow finally agreed, which is how in May of 1970 Spieker came to represent Crow's interests in the Far West. He was a relatively young man, just four years out of college, and now he handled all the developing Crow concerns in some of the nation's most promising markets.

Thus, after brief meetings with Trammell Crow, Allan Hamilton wound up as a real estate developer and Joel Peterson became an agent in France, while Ned Spieker was given one of the largest geographic entities in the Crow constellation. On the whole, Crow's intuitive selections worked out well for the individual and the company both. In that period Crow's people came from a wide variety of backgrounds. For example, Don Williams, who once intended to become a criminal lawyer and who had little development experience, now heads Trammell Crow Company.

Don Williams came from a middle-class New Mexico family. His father, a book and magazine wholesaler, tried to establish a credit-reporting business in Roswell, but wasn't able to make a go of it. He died when Williams was a senior at Abilene Christian College, leaving his son more determined than ever to succeed at whatever he did.

What he did turned out to be law. Williams attended George Washington University Law School on a scholarship, and after graduation considered a career in trial work. In fact, he was offered a fellowship to study for a doctorate in criminal practice, but he rejected it because he had to find a job and make a living. He didn't know where he wanted to live, but he had offers from law firms in New York, Los Angeles, Washington, and Atlanta. Williams and his family decided to go to Dallas, where he found a position at the firm of Geary, Brice & Lewis, working with Joe Stalcup. He and Stalcup left the firm in 1968 to form their own law firm, which performed some legal work for Crow. Williams engaged Crow's interest with his intelligence, diligence and his success in making and retaining relationships, but most of all his character.

There were two groups involved in the operations then: old timers like Gil Thomas, Bob Glaze, and Bill Cooper, and new arrivals such as the young men just discussed. Williams was one of the more impressive of the second group, and Crow decided he wanted to work with him closely. The idea bubbled for a while inside Crow, awaiting the proper time to surface, which so often is the case when it comes to developing properties and taking on new people.

Probably the decision-making process began in the spring of 1970, when Crow called Stalcup about the possibility of his taking a trip to Hong Kong, where Ewell Pope was working a deal to build a hotel. Wanting another assessment of the situation, he asked Stalcup to accompany him to the site, look things over, and provide the benefit of his insights. At the time Stalcup was deeply involved with Crow's work on the Embarcadero Center in

San Francisco, and matters had reached a critical point, so that he could not go. He recommended sending Don Williams in his place.

Crow knew Williams, of course, and had been impressed by the way he conducted himself, but wasn't certain that he was the right person for this particular job. Stalcup, who understood Williams far better than Crow did at the time, assured him that the 29-year-old lawyer could handle matters quite well. One day soon after the conversation with Stalcup, Crow and Williams happened to share an elevator in the Stemmons Tower. After some idle chatter, as casually as he could, Crow asked, "Oh, incidentally, can you go to Hong Kong with me next week?"

He could and did, and he was to spend much of the next decade fanning flames when necessary, and putting out fires when that was required, both on the road and at home. Within two years of the initial trip to Hong Kong he was placed in charge of Crow's international operations, creating a good deal of order out of the chaos in that field, straightening out personnel, getting Crow out of scrapes—in fact, performing many of the duties of a chief operating officer. The more Crow observed his actions, the more they talked and worked together, the more he was impressed with Williams's abilities.

Williams had a special knack for working with people. For example, he saw in Peterson qualities that Peterson himself may not have realized were there. When Peterson thought of leaving the organization he talked it over with Williams. By then the firm was becoming embroiled in financial difficulty (as will be explored in Chapter 7), and Williams, who by then had assumed charge of Crow's international operations, was in the midst of it all. He suggested that Peterson might be interested in returning to Dallas to help him work on these problems. It wasn't what Peterson wanted; he was primarily concerned with being a leasing and development agent, since that was the established path to a partnership. Williams persisted, pointing out that Crow was in no shape to extend itself into any new areas and that the key at that time was in finance and administration. He also promised Peterson that when it was all over—when the company started to expand again—a place would be found for him in development. Even then, Williams had started coordinating operations and was becoming involved in the development of partners.

As will be seen, Williams succeeded in working out of a difficult situation in Hong Kong, for which he was later rewarded with a partnership. Years later Williams remarked, "The amazing thing is that I was ready. I had no idea that I was equipped to handle that kind of responsibility." Perhaps so. But Crow's belief that individuals with character, intelligence, and diligence are capable of more than they might imagine was often vindicated.

Some of Crow's old friends and associates complained that he tended to be too trusting, leaving himself open to being taken advantage of. Bob Kresko said that Crow placed too much faith in too many people and that on occasion he had a tendency to let the stronger people in the company fend for themselves and to protect people of lesser proficiency. Yet on the whole his approach usually seemed to work well.

However, Crow's method of conducting business presented problems. What might have happened if Crow had become disabled in the early 1970s? Glaze might have been able to coordinate matters for a while, but though he was a strong implementor of programs, he lacked the interest and ability Crow possessed for making deals. Most likely in Crow's absence the company would have flown apart in a matter of months, with each partner going his own way. In other words, Crow had not erected a real company or even an organization; there was no institution in place that could survive its founder. Bob Glaze and a few others were in constant anguish over the lack of unifying structure, but if Crow ever gave the matter any serious thought, it was not evident.

Glaze appreciated his chief's spontaneity even though yearning for more stability. Though he might criticize the man and importune him to reform his ways, no one labored harder for Crow than Glaze did in this period of turmoil. One night, returning from a party, Trammell and Margaret drove past the Hartford Building and saw a light on in the office. Crow investigated and found Glaze at his desk, poring over a set of books. Surprised, Crow asked why this late work was necessary, and Glaze treated him to a litany of how much there was to be done and why so much extra working time was needed. Genuinely concerned and shaken, Crow told Glaze he didn't want him to put in such long hours in the future. A few days later, however, he dropped a file on Glaze's desk, casually asking him to take care of the matter. Glaze looked it over and realized that these new tasks would add at least 20 percent to his already full workload. Crow sincerely didn't want his loyal partner to put in those long hours, but at the same time he wanted the jobs to be done. It was just another example of his many contradictions.

But were these truly incongruities? Crow's way was to inculcate in those young men and women the sincere conviction that their interests were identical with Crow's. Then Crow could permit them a great deal of leeway, knowing that the intelligent, dedicated partner would work as Glaze did, not only for personal benefit, but for that of a person he highly admired. In 1985 Joel Peterson, by then one of the most influential figures at the Trammell Crow Company, would tell the other partners, "If you think about it, pre-1975 was the most centralized form we've known. All cash was centralized, control was central, all distribution and ownership decisions were central." Yet there surely was more to the company's tenacity than centralization. In those early years the Crow interests were bound together by force of will if not by structure. The bindings were not readily apparent, but no less real for being invisible.

SIX

Overseas

Americans were not born to frown. It does not suit them. The quintessentially American characteristics are cheerfulness, optimism, and generosity, a general buoyancy of spirit, a belief that tomorrow will dawn a better day.

—THE ECONOMIST, 1987

When Trammell Crow's expansion into Europe and then to other parts of the world began in the early 1970s, it seemed both logical and natural. Trammell and Margaret loved to travel, and as they went through Europe, he couldn't help but think about how some of the concepts developed in Dallas and elsewhere might be applied there.

These were heady times. In the mid-1960s America seemed optimistic and assertive, certain of its values, aspirations, and abilities to accomplish almost anything. In those years foreigners, though they may often have disliked American actions and approaches, nonetheless respected both, and stood in awe of the country's achievements and energy, especially in business.

That Americans should expand commercially into Europe was considered both natural and beneficial for all concerned. "The American Challenge," wrote Jean-Jacques Servain-Schreiber, "is not ruthless, like so many

Europe has known in her history, but it may be more dramatic, for it embraces everything." "Everything" included real estate development. American builders were welcomed abroad because they were strong in the development experience Europeans of the time lacked.

In learning how to operate abroad, Crow and his associates discovered that techniques that worked well in Dallas, Atlanta, and Denver were ineffective in Paris, Milan, and Madrid. This is taken for granted today, but in the early 1970s, when so many people were telling him the world was waiting for Americans to come into their country and show them how it is done, Crow was both certain of his abilities to work deals abroad and confident of success. He arrived, but he was taught quite a few lessons, and he never was able to perform as well there as he originally hoped. Those first years overseas produced an almost unrelieved litany of woe.

In 1970 Crow was visited by two representatives of Investor's Diversified Services (IDS), the Switzerland-based financial operation. Later IDS would be found in violation of the laws of several nations, but at the time was considered an innovative and imaginative operation, and had on its staff such people as James Roosevelt and Pierre Salinger. The IDS spokesmen, Glenn Isaacson and Dick Palmer, were interested in lending Crow money, which made them quite special in his eyes. Crow promptly turned the discussion to the issue of their opening an office for him in Europe. So the following year Isaacson and Palmer joined Crow to help plan such an operation.

Knowing he was entering a strange new world of development with partners he didn't know very well, Crow was more cautious than he had been in any of his earlier expansions out of Dallas. He intended to start small and then, when comfortable in his dealings, go into Europe in a big way.

Initially Crow united with Mack Pogue for individual operations, and a joint Trammell Crow–Lincoln Property office was established in Geneva in 1971. Their first commercial project was in association with Rodman Rockefeller's International Basic Economy Corp. (IBEC), a branch of that family's empire established to invest in developing countries. The Geneva office became the hub for Crow's development in European real estate. Crow wanted to build office buildings and warehouses, while Pogue would erect housing.

Crow began in Spain, with an extension of IBEC's activities for a Dutch business called Makro, which was a wholesale, cash-and-carry food and nonfood distribution outfit—in effect, a supermarket operation selling to smaller "mom and pop" retailers. After several meetings with the IBEC representatives in New York, Makro came to Dallas to talk about expansion in Spain. They agreed on a working relationship and determined the initial store should be in Madrid.

From the first Crow became aware of the need to operate under a different set of rules from those applied in the United States, which in comparison with the rest of the world is quite free and easy. Under Spanish law, for example, foreigners could own only 50 percent of any business. Crow had an old connection who agreed to come in with him as the Spanish partner. So all of them—Cucurella, IBEC, Makro, and Crow—became involved in an ambitious project to create a chain of stores in Spain, the construction of which was to be handled by their Spanish operation, to be known as Espanimmo S.A. Crow had 25 percent of the deal, Isaacson took another 25 percent, while Cucurella nominally had 50 percent, with the others in as lenders. In addition, there was an understanding that Don Williams, who would oversee matters for Crow, would have half of Cucurella's share.

Financing in Spain was particularly difficult because the Spaniards had a different notion from Americans of what constituted a long-term commitment: To them it was six months or less. Crow learned that the same rules that pertained to ownership went for borrowing, in that foreigners had to bring half their financing from overseas. This meant he could not borrow sufficient pesetas to finance the project. So Crow arranged a split loan between a Spanish and a Swiss bank, the former in pesetas, the latter in dollars to be converted into pesetas. These were five-year loans, long-term in their view, very short from Crow's way of thinking, but there was nothing he could do about it.

The partners put up a large Makro store that was later sold at a profit, and the experience encouraged them to go further. Espanimmo, as the Madrid operation was known, erected a similar warehouse for Makro in Barcelona, and then developed a condominium and office building as well. At about the same time they purchased a parcel of land in Madrid to build an office building—and lost enough on that deal to use up all the profits generated by the food stores and a few million more.

There were persistent personnel problems. Glenn Isaacson seemed to be performing well, but he was more interested in apartments than in commercial construction, and so spent more time on Lincoln's business than on Crow's commercial ventures. This neglect was taken care of in 1972 when Isaacson introduced Crow to his brother Larry, who at the time was teaching at Harvard and acting as a consultant for S&H Green Stamps, and who was interested in commercial real estate development. Larry Isaacson came into the Crow organization in late 1972, and he was to play an important role in its European commercial operations.

In many of his European projects Crow dealt with English real estate advisors, known there as chartered surveyors. These are supposedly detached, competent people who among other things appraise market conditions, package real estate projects, negotiate construction contracts, have a hand in the design, help out with financing, and handle myriad other related tasks, after which they bring the entire package to potential investors. That was the traditional way of entering into real estate deals in the United Kingdom, and it had started to be common on the continent as well.

One of these chartered surveyors, a firm by the name of Weatherall, Green and Smith, brought Crow a package for an office building in the western suburbs of Paris, a small town called Louveciennes. The project looked solid, so Crow and his associates decided to make this their initial undertaking in France.

Work began in 1972, and immediately they encountered difficulties seldom experienced in America, and certainly not in Dallas. Neighbors filed a suit to cancel the building permit because the structure would block their view of a park across the street, and they won. So halfway through construction the permit was canceled. The partners applied for the reinstitution of the permit while continuing with the construction, knowing this was risky, since a second defeat in the courts would mean the building would never be occupied and would have to be razed.

The case dragged on, the building was completed, and it sat vacant for 18 months while the partners waited for a court decision. They eventually won in the French superior courts, and were finally able to offer leases. Along the way Crow had to sue the contractor who had guaranteed the project and all conditions (as well as his predecessor), but the contractor went broke. Eventually Crow sold the building to a French investor, and emerged from the ordeal with a small loss.

Soon after the partners became involved in the Louveciennes project, Weatherall, Green and Smith called their attention to another venture, a large office building on the east side of Paris, in the commune of Bagnolet. This was a large twin tower enterprise, for which the French bank Credit

Lyonnais was willing to make a commitment to a fixed-rate, 15-year, interest-only mortgage, something unheard of in France at that time. The loan was guaranteed in part by two British pension funds, which would own 40 percent of the project. This project went more smoothly than the first French one, but Crow ran into a familiar problem when the work was finished. The property came to market at a time when Paris was experiencing a commercial real estate glut, and Crow had serious difficulties in placing leases, obliging him to offer concessions. The pension funds changed their policy and decided to end all joint ventures. They agreed to buy out the partners for $2 million, which was roughly what had been put into the operation.

By then the Isaacsons had gotten the partners involved in other deals elsewhere in Europe that were turning sour, and Crow started drawing away from them. In 1973 he had persuaded Don Williams to join the company to oversee the international operations. Williams promptly set about solving problems, which involved making personnel and other changes necessary to bring stability and profitability to the operations. One change was the selling of the properties in France, which seemed sensible considering that the company didn't really understand the market there.

Crow's experiences in Germany were better. In 1972 Glenn Isaacson hired Karl Homberg to initiate a joint housing and commercial property development program there. Shortly thereafter, however, Isaacson lost confidence in Homberg and recommended his replacement. By then Williams was regularly shuttling back and forth between Dallas and points in Europe and Asia. He looked into the situation and concluded that Homberg was doing a good job and that significant development opportunities were available, specifically in developing warehouses and offices in Dusseldorf and Cologne. It was unavoidable that Williams's opinions would cause frictions with Isaacson, but the fact remained that Crow's German operations did better than those elsewhere in Europe, and continue to do so.

The next stop was Italy. In 1973 Crow opened an office in Milan, under the leadership of Jack Poulson, an American expatriate with a good knowledge of the country. Poulson built a warehouse in Milan whose construction presented Crow with the same kind of red tape and difficulties encountered in France.

It was dawning on Williams that their problems stemmed from a lack of leadership in the European offices and the inability to master the art of conducting business in the European arena. The kind of relationships Crow had enjoyed in the United States, based on mutual respect and honor, were more difficult to develop in Europe, due to significant cultural differences. For example, the Madrid office building deal was done in partnership with two developers who had assured Crow that they had arranged all the financing and there would be no trouble with permits. After the Isaacsons and Crow entered the transaction it became clear the local partners would be unable to execute their end of the deal, and they had to bring in others to take over. Williams then brought in Larry O'Brien and Jose Pelayo to open an office in Madrid and try to bail the partners out as painlessly as possible. On top of all this, Jose Maria Cucurella asked to be let out of the deal. Eventually Pelayo was able to sell the land in a soft market, at a loss of around $5 million. And finally, when they sold the Espanimmo company, Cucurella attempted to withhold Williams's share, and they had to file suit against him to obtain the funds.

Williams was obliged to spend an increasing amount of time in Europe, at a time when he was also needed elsewhere. He managed to stanch wounds and cut back on losses while winding down most of the unappealing operations. Perhaps Crow should have pulled out of Europe in 1973. But the partners had so many major projects underway that it wasn't easy to alter course.

There were several important personnel changes, however. As a result of the troubles Crow was having with Larry Isaacson, Williams first cut back on his responsibilities and then, in 1974, informed him there really wasn't much of a future for him at the firm. Isaacson left soon after. By then they had brought in some new people, many of them native to their countries of operation, and while their results were better, difficulties remained.

Joel Peterson, who was in Lyons in 1973, had been asked to evaluate the French projects and operations and recommend major changes. Williams engaged a new manager for the French operations, Phillipe Vergely, a capable, experienced French developer, who completed and then disposed of the Louveciennes and Bagnolet projects as painlessly as possible. Another Frenchman, the aforementioned Gil Rougier, was placed in charge of the Lyons projects and given a mandate to enlarge the commercial business in the south of France, doing so with mixed results. Rougier and Peterson did create a major warehouse project in Lyons. It was successful in terms of acceptance but eventually lost money. At the time the warehouse was launched, it was not possible to obtain 100-percent French franc financing, due to changes in French exchange regulations. So the partners arranged dual financing, partly in francs, the rest in a basket of Euro currencies (known as a "currency cocktail") that fluctuated sharply against the franc—and against the partners as well. They sold the project for a French franc profit of around $1 million, but that much and more was lost in currency translations. Vergely and Rougier have since continued the Lyon operation, have been successful, and today are major French developers.

Looking at the whole European picture, it might seem that all Crow had was a string of blunders and losses with a few successes interspersed among them that lured him deeper into the arena. Yet some of the ventures developed nicely. During this period Dick Palmer and Glenn Isaacson had a falling out, and Williams moved Palmer from Geneva to Brussels to handle affairs there, starting out with something Crow knew very well: warehouses. This program was quite successful, but in the midst of it tragedy struck. In 1977 Palmer suffered a serious, debilitating stroke. Because of the way Crow conducted business, without structure and personnel training programs, all depended on the individual. When Palmer left the scene, the business prospects disappeared. So all the Brussels buildings but one were sold, and Crow closed his office there. Even when Crow won, he lost.

Marts were one of Crow's primary interests in Europe. From the first he intended to construct a string of them, at least one in each European country. This was no wildly ambitious idea of thrusting a replica of the Dallas Trade Mart upon an unwilling market, but rather a synthesis of long European tradition and American drive. As noted in Chapter 3, trade fairs dated back to medieval times and were deeply ingrained in the social, commercial, and political fabric of the European countries. Bringing in John Portman, Crow planned to go into the major marketing cities of Europe—Frankfurt, London, Brussels, Antwerp, Paris, Milan—all the cities that hosted major fairs each year. He intended to acquire land close to the existing fair ground in each city and to erect a market building, which he thought would be quickly accepted by the manufacturers and distributors in those areas.

Each country presented different problems due to regulations, patterns of conducting business, and values and preconceptions of those involved. Obviously, any American businessperson attempting to operate abroad must devote much time and care to appreciating the different mores, knowing the tuition for such an instructive course is bound to be high.

Crow and Williams recall a luncheon meeting with a group of British bankers in 1971. Crow arrived on time, and his hosts were well prepared and genial. To Crow they looked like extras in British movies, with their black suits and ties, and white, starched, pointed-collar shirts. His own attire probably appeared ordinary enough, but like him, they might have seen an image based on something else—a cowboy riding out of a Western film into their midst.

The luncheon was quite pleasant. They talked of many things except the matter at hand, which Crow thought best to leave to his hosts to raise. When the meal was over, one of the bankers turned to him with a quizzical but friendly smile. "Well, Mr. Crow," he said. "You have passed the first test for doing business with Schroeder's Bank. You didn't attempt to bring up your business at luncheon. We have a policy dating back to the 19th century that if anyone begins discussing a proposal at the first luncheon, it means that he needs money too badly for the investment to be a safe one for us. So won't you now tell us something about yourself, and what you have in mind?"

Crow looked around the table at the assembled bankers, realizing that everything was going well. "Gentlemen," he announced, "we're a large private developer in the United States. You know, real estate developers where I come from are about 50 percent con men. The difference between me and the others is that I am only about 20 percent con."

They roared. The ice was broken. But little was accomplished with this group, due to complications and delays that were encountered. Such encumbrances were quite familiar to British developers, but not to Crow, who was accustomed to a less constrained business atmosphere.

The constraint was frustrating for Crow, who had concluded that if any city outside the United States needed a mart then, it was London. As one of the world's great commercial centers, the British capital hosts hundreds of business conventions of various kinds each year. Yet then (and even now, for that matter) it lacked the kind of exhibition facilities such a city might have been expected to have.

Crow and Portman planned a large, elaborate mart-hotel complex near London. In 1969 they set out to find a site. The most obvious location would be near Heathrow Airport, so that businesspeople could fly in and out for their meetings. Locating the land presented no difficulties; there was plenty of open space near Heathrow. Obtaining necessary approval for the mart was another matter. Crow and Portman wended their way through various commissions and citizens' groups for two years, only to learn that there would be no further development permitted in that area.

So they located another place, this time a run-down area near the Surrey Docks on the Thames River in London. It wasn't as convenient for the air age as Heathrow, but it was not devoid of appeal. Businesspeople would be near other contacts in London, there was the entertainment value of being in one of the world's great cities, and Crow could concentrate on the mart and, initially at least, not have to be concerned with hotels.

He did not anticipate much trouble. After all, he intended to replace an unproductive eyesore with an attractive and useful structure. But Crow underwent a repeat of the Heathrow experience. Again he made his case at the various governmental levels, from a local council to a regional council, all the way up to Parliament itself. Crow went through the drill for three frustrating years, and toward the end the land was nationalized and placed under the control of the Greater London Council and the Suffolk Borough Council. Negotiations had to begin all over again. In May, 1976, Crow finally concluded an agreement with the new landlords, but by then England was in the grips of stagflation, with interest rates on long-term debt at 18 percent, and Crow couldn't build in that kind of environment. At this point one of the leaders of the London council suggested that the money for the mart be raised through Parliament, but this idea foundered. The council was Tory, the government was Labour, and the bill was defeated in 1978.

London never did get that mart, but the partners did erect one in Brussels. In fact, they had started on that one first.

Brussels has a population of 2 million and is noted for its fine architecture and, in more recent years, its status as the headquarters for NATO. It is one of the continent's oldest cities and has been a commercial center ever since the Middle Ages. In the post-World War II period Brussels became an entrepot for many American firms, which located their overseas headquarters there.

Crow was not unfamiliar with Brussels. One of his partners, Maurice Moore of Continental Trailways, a company in which Crow was a major shareholder, purchased engines for his buses from a Brussels supplier and had some good contacts there. So in several ways it was a natural place for Crow to consider erecting a trade mart.

Crow first considered Brussels as a mart site in 1958, when he visited the World's Fair there. While recognizing that the city was a superb location for a mart, he was put off by Europeans who informed him that the reason there was no such facility in Europe was because their businesspeople preferred the kinds of temporary fairs that had been utilized for centuries. He persisted. Actual negotiations for the Brussels property began in 1969, when Crow joined with a group of Belgian investors and approached the government with a set of preliminary plans. The following year Brussels obtained title to part of the site of the World's Fair, and with this discussions became more intense. Initially it appeared there would be few problems. The Belgian government was prepared to provide a quarter of the capital, for which it received 25-percent equity ownership and the right to have a representative on the board.

Yet nothing went forward, due to the unwillingness of the Belgians to finalize agreements with the speed Crow was accustomed to finding in most parts of the United States. He returned to the issue in the early 1970s, while negotiating for that British facility. At that time he entered into an alliance with Maurice Naessens, president of the Banque de Paris et des Pays-Bas, and it was then that Portman was brought into the project. Portman went on to design what became the Brussels International Trade Mart. As usual, his plan was magnificent, and his cost estimates too low. Inflation in Belgium (and much of the rest of the world) compounded the difficulties.

Costs were only part of the problem; the project had complications from the first. Government officials proved difficult to get along with, probably because of cultural differences. They were far more conservative than the Americans, and often held them back. The banks were impossible to deal with. Their leaders would change their minds—and pledges—from day to day. Construction finally began in 1973, and the mart was completed two years later. The structure contained 1.5 million square feet on four levels, and could be expanded without much difficulty to 3.5 million. Eventually it would serve several industries, including furniture, gifts, jewelry, apparel, toys, and sports equipment among others.

The primary market for the Brussels facility was northern France, western Germany, and the Benelux countries, while the secondary market was the rest of the Common Market. Once a year Crow ran a furniture show that brought in buyers and sellers from all over the world. In addition, there were several "market days," during which all the 1,200 exhibitors are on the floor, and buyers for large department stores and chains can place orders for various products at the same location.

Business at the Brussels Mart started out slowly. In part because of concern over this, one of Crow's American investors in the project, Joel S.

Ehrenkranz of Ehrenkranz, Ehrenkranz & Schultz, New York-based attorneys specializing in income and estate tax matters, went on its board. Ehrenkranz had invested client funds in Lincoln Properties projects as early as 1969 and met Crow through Mack Pogue in 1971, when he was told of plans for the Brussels project. He was introduced to Portman, and apprised of the details of the deal. Ehrenkranz's investments in the mart, and so his presence there was indicated. Thus began a string of Ehrenkranz participation in Crow deals, as well as a close friendship.

Within a few years the Brussels Mart became an important marketing vehicle in western Europe, and by the early 1980s it turned the corner on profitability. Even so, the facility was not what Crow had intended it to be: one of a string of marts stretching across western Europe.

Looking back, Crow realized he made three mistakes in his European operations. In the first place, he tried to expand too rapidly. The pace of business in most European countries is slower and more relaxed than it is in the United States, a fact that Crow did not sufficiently take into account. Like far too many American businesspeople of the time, he was aware of cultural differences among American cities, such as Atlanta and Los Angeles, and adjusted his approaches when dealing with counterparts in those cities, but he didn't fathom the far deeper differences among European cities, and he suffered because of his lack of understanding.

In the second place, Crow lacked the proper grounding in financing and currency risks. He acted as though a bank was a bank and an insurance company an insurance company no matter where it was located, and of course this simply isn't so. To be sure, currency risks were minor in the late 1960s, but gyrations became the norm after President Nixon took the United States off the gold bullion standard in 1971, and commitments made at a time when such considerations were minor were now at grave risk.

Finally, and perhaps most importantly, when Crow became aware of his shortcomings and tried to overcome some of them by uniting with native businesspeople in each country in which he operated, taking them on as partners, he chose poorly and established the wrong ground rules. The trouble was that the Europeans were not partners in the same way as American partners were. As Crow put it, "They were traveling on our nickel. If all went well, they would do well, but if a particular deal turned sour, they went home free, and we would be stuck with a failed deal."

Crow never totaled his European gains and losses in the 1970s, but not counting Germany he definitely suffered a net loss. Simply stated, the real estate risks, the currency risks, the legal risks, the political risks, the management risks—among others—made American real estate operations in Europe very dangerous. It took Crow some time to learn that there wasn't enough potential profit to justify all those hazards, not when he could do just as well or better, with less exposure, elsewhere.

Crow fared somewhat better in Latin America than in Europe, although the experiences there were not overwhelmingly successful. Makro was pleased with his work in Spain, and in 1972 Crow and Williams traveled to São Paulo, Brazil to discuss putting up a building for the company there. By then Crow and his associates had learned the importance of having the right local partners, and they found one in the Esteve family, which was related to the Cucurellas. The deal was financed by the First National Bank of Boston, which provided 80 percent of the requirements in the form of a 12-year loan in dollars. It was next to impossible then to get long-term financing in Brazilian cruzieros, due in part to tradition but more importantly to the bouts of soaring inflation that were endemic to Brazil. The Brazilians found their solution to the problems of inflation through indexation, which meant that rents were tied to fluctuations in the value of their money. They quite regularly devalued their currency in relation to the dollar, but indexation tended to protect foreign investors against losses, so devaluation wasn't a major problem.

Crow's biggest problem was trying to run the deal from Dallas. He soon realized he needed to establish a Brazilian office headed by a local businessman. Williams assumed the assignment of locating such a person, and after much searching settled on Peter Justiniano, a Chilean employed by W. R. Grace in Brazil. Justiniano came to Dallas to meet Crow, who agreed that he should have the job. He did well with the Makro project in São Paulo, and also on a second store erected for Makro in Rio de Janeiro, backed by a 12-year loan from the Marine Midland Bank. He then erected a 180,000-square-foot hypermarket, really a large discount store, the first in Brazil. Justiniano also put up a store in São Paulo for a Dutch group known as SHV. This project was done in partnership with a Brazilian investment bank called Crefisul. All these undertakings were successful, as were several others that followed; the hypermarket was sold for a substantial profit. But economic problems in Brazil, exacerbated by shaky finances, led Crow to slow down operations in that country. Justiniano remained and remains his Brazilian partner, however, and continued to bring in interesting joint ventures with limited risks but potential profits through ownership interests.

Crow had some personnel problems in Brazil, though not with Justiniano. It was yet another case of obstacles presented by cultural differences between American and foreign businesspeople. Crow had a falling out with both the Esteves in Brazil at about the same time he was breaking with the Cucurellas in Spain. In planning the hypermarket, Crow had brought in the Esteves for a 25-percent share. They dragged their feet about signing up, waiting for Crow to put everything in order before coming on board, when all the risks had been eliminated. At that point the family swept grandiosely into Dallas as though nothing was wrong. Crow did not admit the Esteves into the partnership in that deal, and he never again worked with them.

The early experiences in Asia and Oceania also had mixed results. As mentioned before, Ewall Pope correctly anticipated the great Hong Kong business and land boom. He spoke of it with Crow, and they agreed to form a new entity, Crow-Pope, which would erect a hotel there. Crow-Pope, like so many Crow enterprises, was not meant to be a fully functioning company engaged in a wide variety of businesses. Its sole concern would be that Hong Kong hotel, and the creation of such an entity—whose total capital consisted of $40—was meant to limit the partners' liabilities.

The first step was to secure a site. Pope traveled to the Crown Colony and found what he considered the best location there for a hotel. In seeking out its owner, he learned that land was scheduled to be auctioned. He telephoned Crow to discuss just how high they would go in the bidding. Several years later Pope recalled telling Crow, "Trammell, we've got to start off basically at this price and we've only got $40 in the company, if we should get the bid. We have to pay 10 million Hong Kong dollars [approximately $1,870,000] for a long-term lease." Pope then added that under the law, 10 percent of the bid would have to be paid within 30 days for what amounted to an option on the property. This meant the partners would have to raise $180,700, which they stood to lose if the project did not get off the ground. Pope urged Crow to go along with such a deal. "As I see it, the worst that can happen to us is that I lose half of that $180,700. You lose half of that, and we lose that $40-company we established so fast." And there always was the possibility of selling the option to someone else at a profit.

The bidding went higher and higher, and Pope stayed in the race, taking the property for approximately $21 million, far more than had been anticipated. Pope sent Crow a message: "You can't believe what I had to pay for the hotel site." And he was right—Crow was stunned. Pope had paid $550 a square foot, one of the highest land prices paid in the world to that date. There was no way to call the bid back; Pope clearly had lost his head, and Crow did what he could to make the deal work. This was when Crow called in Joe Stalcup, and wound up with Eugene Locke and Don Williams.

By then a senior partner at Stalcup's law firm, Locke had recently served as deputy ambassador to South Vietnam, and he knew the Far East well. On his return to Dallas he had become one of Crow's key negotiators. Locke and Williams had met before but had not yet worked as a team. When they arrived in Hong Kong they learned from Pope that a short-term loan had been arranged with First Citibank to cover the land purchase. Their job was to arrange long-term financing and to engage a major hotel chain to assume the development, ownership, and management of the proposed hotel. This they did, lining up Hutchinson International as the builder and ITT Sheraton as the manager, with the financing to be done by Citibank.

Within a matter of weeks, however, the deal began to unravel, and Crow sent Williams back to Hong Kong with a general power of attorney from Pope and himself to complete the deal. Convinced the errand would be routine, Williams made the trip with his father-in-law, intending to stop off in Thailand afterward for a short vacation. When he arrived in Hong Kong, however, he found that the deal had blown up. The principals were all at each other's throats. Deciding that salvage was possible only if they were separated and negotiated with individually, he did that. For the next three weeks, day and night, Williams shuttled back and forth between his hotel and their offices, slowly piecing the project together again. He had difficulties keeping in touch with Pope and Crow, so he had to make critical decisions on his own.

Out of the transactions came the Hong Kong Sheraton, which later was sold to a Hong Kong investment group for a substantial profit. It was a success from the first, justifying Pope's risks and demonstrating Williams's abilities. And, of course, the adventure put Williams on the road to a partnership, which followed soon after.

As the gamble in Hong Kong showed clear signs of paying off, Pope succumbed to the temptation of projecting Crow, Pope & Land more deeply into the Far East. Jim Hawes, who was named the Crow, Pope & Land representative in Hong Kong, was instructed to keep an eye open for possible deals, and in 1972 Crow himself took an extensive trip throughout the area, stopping in Japan, Hong Kong, Thailand, Singapore, Indonesia, and Fiji to check out recommendations Hawes had forwarded to him. After each stopover Crow would write to Williams, telling him what he had found and in some cases asking him to follow up. It was a typical Crow foray overseas: much talk, many dreams, and then complication. Only this time, unlike on some of the sallies into Europe, Crow was more circumspect.

Indonesia, in the midst of a boom, seemed the most promising area. Through mutual friends Crow met Jan Dharmadi, a Chinese who had married an Indonesian and had changed his name to appear native-born, since there was much antipathy toward the overseas Chinese in Jakarta. Dharmadi owned several gambling casinos and a high-rise massage parlor, and seemed well on his way to becoming a local magnate.

For a while Crow and Pope considered using Dharmadi's services in putting up a hotel and office building in Jakarta, but the deal foundered due to another of those stumbling blocks Crow always seemed to find when venturing overseas. In most parts of the United States there is little difficulty getting a title search and discovering exactly who owns a parcel of land. Not so in Indonesia, where there were all kinds and grades of land titles. Crow also encountered the now-familiar problems of procuring local financing. Those capable of arranging financing charged high prices for their services, and often were unreliable. Foreigners attempting to go it alone in Indonesia soon learned deals were impossible without such assistance. In the end Crow discarded the idea entirely and the hotel–office building was never constructed.

Then on to Fiji, where Crow met Bill Dailey, a Harvard M.B.A. who became a photographer and dropped out of mainstream society in a way that had become fashionable in the 1960s. He wound up in Fiji, in partnership with an Indian who had become the largest owner and operator of taxicabs on the island. Together they accumulated land, and were seeking interested builders. Again, as in Indonesia, there were problems with land titles. Crow asked Williams to take a hand in the dealings, and after some investigation, the younger man advised against attempting anything in Fiji. So Crow moved on.

By then Crow was mulling over an idea presented by another partner, Don Russell, who was intrigued with the idea of creating a string of hotels in the South Pacific. The Japanese had recovered from their wartime experiences, both financially and psychologically, and were becoming very interested in tourism. The surge could already be seen in Hawaii, and doubtless other areas would follow. In addition Americans and others were turning to the Pacific areas, which to Russell seemed the last great frontier for development.

Russell took his ideas to Intercontinental Hotels, a subsidiary of Pan American World Airways. He proposed to develop and build hotels that Intercontinental, with its expertise in management, would operate. On the surface it made sense for all concerned; Crow and Russell entertained visions of Pan American flying tourists to Pacific hotels constructed through Trammell Crow and operated by Intercontinental. This vision set the stage for one of Crow's more interesting, and in a way bizarre, experiences.

The location selected for the first hotel was Port Vila in the New Hebrides, a tiny island chain northeast of Queensland, Australia. First charted by Captain Cook in 1774, the islands today make up the nation of Vanuatu, but three decades ago they were under the joint administration of the United Kingdom and France. Since the two nations constantly squabbled and seldom agreed on anything, it had two bureaucracies, two sets of laws, and almost no effective system of taxation—which made New Hebrides an excellent tax haven. About 130,000 people lived there, most of them Melanesian, who have adopted pidgin English as their commercial language.

It is hard to imagine a setting more romantic, attractive—or remote. Port Vila, the country's commercial hub and capital, had a population of 15,000 at the time. The economy was based on fishing and copra, and the interior of New Hebrides was a thick, almost impenetrable forest. Crow had considerable difficulties getting the hotel erected because of problems in transporting materials and crews to the New Hebrides. And there were problems in managing the hotel once it was up.

The hotel does a good business now. As it turned out, most of the tourists come from Australia, not Japan. So whether the company likes it or not, it owns a romantic hotel in the South Pacific, the unlikeliest jewel in the crown. Crow himself is embarrassed to talk about it. "It's a very nice hotel with a beautiful lagoon, but it literally is at the end of the world," he wistfully remarks. "It is so hard to get there that few people go. To travel to Port Vila you have to spend a night in Fiji on the way out and another one on the way back, and possibly another night in Hawaii to make airline connections." Pan American did not put into operation the direct flight he had expected, because the business was too small to make it worthwhile. And, making matters worse, in recent years Vanuatu has become closely affiliated with the Soviet Union.

Williams and Crow took an extended trip to Hong Kong and Taiwan in 1973 to seek out additional prospects. At the time they had a Taipei representative for Market Center business, John Ni, through whom they hoped to transact some business. They soon learned that difficulties obtaining financing would prevent any successful dealings there. The Crow representative in Tokyo was Ellen Cooke, a Harvard M.B.A., who was somewhat of an anomaly since the Japanese businessmen were not accustomed to conducting business with women. Like Ni, she had little initial success. But Crow retained hopes for significant American investment in those days when it seemed the Japanese were mere imitators who had much to learn from Americans, so they persisted, in vain.

Williams also made a foray into the Middle East, taking several extended trips to Beirut, Cairo, Jiddah, Kuwait, and Teheran. An attempt to purchase land on the Nile for a hotel didn't work out, and after doing quite a bit of work for a bank building in Teheran, he was informed that his would-be partners would not be doing business with Crow. All they really wanted to do was pick his brains.

It probably was all to the good that Crow did not become more deeply involved overseas than he did, especially outside of Europe, although later on the company's Pacific ventures became more promising. The most important reason for not expanding energetically overseas in this period was that all of Crow's energies and assets were needed to meet the great real estate crisis that was brewing in America in the mid-1970s.

SEVEN

The Crunch

> *I once heard it said that the cat that is burned on an oven range will never touch a hot one again. True enough, but that cat won't go near cold ovens either. The same is true for business. Failures that transform a businessman into a super-cautious individual can cripple, and this attitude has to be guarded against. I don't watch my pennies. But I don't deal in pennies either.*
>
> —TRAMMELL CROW, 1980

The beginning and the end of great historic periods are apparent only in retrospect. Occasionally perceptive observers may deduce the shape of transformations while they are occurring. Even then, however, it is difficult to develop sound strategies and tactics which will enable a person or company to adjust to the changing times. The past can provide a compass, but not a road map.

As mentioned in other contexts in previous chapters, the United States, and indeed the world, underwent major changes during the late 1960s and 1970s. One age was ending and a new one in national, perhaps world, history was beginning. It was a time of tumultuous social and economic transformation such as hadn't been seen in the post–World War II years;

in fact, somewhere along the way the "postwar period" ended, which is to suggest that the great conflict of the 1940s no longer was a basic reference point. An age characterized by optimism, confidence, and a conviction that progress was inevitable in "the American Century" came to an end, and the country entered a new era, the outlines of which are only now becoming visible.

Probably it is too early to say just when and why the transformation happened, because there was no single factor to which to attribute the changes. What is certain, however, is that the maxims that directed the activities of American businesses weren't working as well as they once did, and in some cases they proved utterly inappropriate.

Along with so many of the nation's institutions and businesses, real estate development was changing, and Trammell Crow was not alone in failing to comprehend this in time to make necessary alterations in course. Had Crow discerned the shifting pattern in the late 1960s, his activities in the early 1970s would have been quite different from what they were.

Of course, businesspeople in any epoch can look back and see where they should have gone, given the proper crystal ball or guru. However, for Crow there was something far different and more serious. He had been able to thrive in that joyous and optimistic period because he had the necessary outlook and talents for such times. It would be different in the subsequent years, when optimism could be a fatal attribute and caution and prudence were far more appropriate.

Crow's difficulties in the early 1970s derived from two sources: the change in the economic climate and the structure of his operations. Without the former, flaws in the latter might have been overcome or simply ignored. Given the stormy economic situation, however, weaknesses in his business operation proved almost fatal.

Real estate developers consider the ideal situation to be one in which interest rates are low and economic activity high. This implies a business climate in which they have no trouble obtaining low-cost loans for construction and the accumulation of land, and are required to pay little to service their considerable debts, while demands for commercial buildings are high due to a strong economy. Their basic approach to lending institutions is to borrow from many and hope none of them discovers how much they owe to the others. Thus aggressive developers leverage their operations to the utmost.

Naturally, they are distressed when opposite conditions prevail—high interest rates and low demand. Developers have learned to expect such periods and try to ride them out. There was a severe shake-out in 1969, when rates rose sharply. All the Crow projects that were then financed by short-term loans tied to the cost of the money market suffered.

As usual, Bob Glaze was troubled by Crow's free-wheeling ways. He attempted to place him in what Crow considered a financial straitjacket. Glaze would tell Crow that for the coming fiscal year his operations would need, say, $27 million in cash, and that if he couldn't raise that amount cutbacks in plans would be necessary. Crow would look the figures over, and say, "Yes, but this and that aren't going to happen and this will happen. The budget doesn't mean anything. We're not going to do that, and we are going to do this." Glaze would shake his head, argue a bit more, and then walk away. They would strap things together somehow.

The situation in 1969 was particularly severe in the residential part of the business. By itself Lincoln Property experienced cost overruns on the magnitude of $25 million in a soft market. There was talk of liquidation, possibly bankruptcy. Crow and Pogue were able to survive, and when the slump ended they returned to a high level of profitability. The experience was all to the good, but perhaps served both men poorly. They had come through the cycle and were going on to new heights. This may have led them to believe results would be just as fortuitous the next time.

A similar, though more turbulent period developed in 1973 and 1974. The major problem was stagflation, the combination of inflation and recession some economists used to believe couldn't occur. Actually there had been periods when the two had appeared simultaneously. In 1958, for example, the economy was weak and the real estate industry was in one of its cyclical slumps while prices rose fairly sharply. But that falloff was mild compared to what the nation was to experience in the mid-1970s. Almost everyone ignored the warning signs: the energy crisis, widening environmental controls, tight money, a federal moratorium on subsidized housing, the likely impact of imminent federal tax reforms, and hyperspeculation in land. Any combination of these factors could cripple the real estate industry, bringing development to a virtual halt.

In retrospect it is easy to see that Crow should have eased up on development and the accumulation of land, and in general prepared for more difficult times. To do that, however, would have been contrary to his nature.

Glaze constantly asked him to straighten out his financial situation, to cut back on risks. Given his accounting background, Crow should have understood and given careful consideration to this advice. He didn't, preferring instead to plow ahead in new areas of opportunity and challenge. So Crow was setting the stage for the distress he would soon experience, which in turn would lead to the kind of restructuring he should have considered earlier.

Crow had never exhibited the kind of fiscal prudence Glaze might have preferred; his all-out approach had served him well since his Navy days, and he naturally clung to it. He was a free-wheeling developer who liked to work with others of like mind. Even then he would chide Pope that Pogue had accumulated more land than he, and then turn around and gently scold Pogue, mentioning the large inventory Pope had put together. And one didn't know the extent of risk the other faced.

There still was no single corporate voice to speak for the collective interests, no real Trammell Crow Company. While each person knew his own projects, the only ones in the organization who understood all of it were Hope Hamilton, Barbara Collins, Bob Glaze, and of course Crow himself. Each of the partners in the field worked with Crow on his own, having little clear idea of how the other parts of his empire were faring, but suspecting that all was not well. Tom Shutt, for example, said that in those years "we were still a shoestring operation," and "really couldn't afford [to arrange financing for] office buildings." Of course he wasn't referring to Trammell Crow, but rather the partnership with Kresko in St. Louis, which in fact was only a small part of the empire. As far as he was concerned, business was fine. But then, Shutt would have been thinking of warehouses in several parts of the country, and little else. For him, and for most other partners, the Crow interests were not united. They were not all in the same boat, as it were; rather, they were in an armada of boats, some quite small, others immense, with no clear relationship to one another. If one part of the fleet—say, Baker-Crow—sank, the other segments might not be directly affected. Even if Trammell Crow himself were forced to personal liquidation, they could have survived; all that would happen is that the creditors would take over the Crow interests in their projects. Moreover, not all the projects would have been affected, for Crow's downfall would not have affected those partnerships with the Crow children. It was, as Shutt put it, not a company at all, but rather "an amalgamation of an association of autonomous operations."

So the Crow interests were not prepared for major problems because of lack of information and coordination. The partners operated independently out of 13 cities, with more being added all the time. They still didn't know each other well, though there were annual meetings and occasional conversations.

Yet it was becoming increasingly difficult to function in this unstructured fashion. It was one thing to forge success in a casual atmosphere in Dallas in the early years, but it became more arduous as Crow's interests spread throughout the nation. As his troubles mounted, Crow conceded that the system (or lack of same) that had served him so well in the past might have to be altered. Yet nothing would be done until he was obliged to act by the pressure of forces he could not control.

The various Crow businesses were performing well as late as the summer of 1973, and the outlook appeared promising. In 1972 Crow had completed work on 2001 Bryan Tower. By then the first stage of the Embarcadero had opened, plans were being made for expansion at The Trade Mart, and there were indications that the European business would improve.

In 1973 Crow initiated $400 million worth of projects. That year the various Crow interests were worth almost $1.5 billion, and Crow's net worth came to some $70 million, with various family trusts adding another $45 million. The Crow partners shared this success. Thomas, Shutt, and Glaze each were worth several millions, and a dozen other partners were millionaires. All this success had come from borrowing heavily and taking chances in an economic environment that rewarded risk-takers. To finance his highly leveraged operations—enormous debts, relatively little equity—Crow had signed notes for more than $500 million for his approximately 650 individual companies.

Undeterred by signs of a weakening market, Crow was, as always, engaged in many projects simultaneously. In addition to the expansion overseas and in the residential area, he was developing warehouses and office buildings in Houston. The major Crow project there was an ambitious redevelopment of the ailing downtown of that city. Much of the area was a grid composed of small blocks and wide streets, but one corner contained some buildings and irregular streets. Crow purchased seven blocks there from some 40 sellers, and closed off the streets and widened some of them. The project was financed at $20 million by Metropolitan Life and was undertaken in partnership with Domenick Paino, formerly with Coldwell Banker's San Francisco office, who was recruited for the job. Paino started work on what became the Allen Center, named after the Allen brothers who had been instrumental in the founding of the city of Houston. Crow had in mind a version of the Peachtree Center and the Embarcadero Center in San Francisco, which is to say it was to be a complex of office buildings and hotels that would revitalize the downtown area.

Growth, always growth. Glaze would ask Crow to sell some of the properties to place himself in a more liquid position, and more often than not Crow would refuse to do so. He rolled ahead, erecting acres of new housing, scores of shopping centers, hundreds of warehouses, and dozens of office towers. He was like an automobile without a reverse gear, or for that matter, without brakes. Exhilaration. Fear. For almost two decades he relished the former. Now Crow was about to experience the latter.

While this was a heady period for his interests and partners, a Crow balance sheet—if one could have been drawn up for his many enterprises—would have been rather alarming. More than 20 percent of total assets was in the form of undeveloped land, and carrying costs took a staggering 150 percent of cash flow. That there were weaknesses in the Crow empire should have been manifest, and Crow might have recognized them. As always, however, he counted on a strong market to bail him out of any difficulties.

The initial problems surfaced in the residential area and in the matter of land and apartment inventories. Baker-Crow had enlarged its land purchases for two reasons. First construction activities were expanding, so more land was needed. The partners also were caught up in the whirlwind of rising prices. Like so many others, Baker-Crow attempted to acquire sites before their prices rose even higher, and so did other partnerships in which Crow was a participant. In time the various Crow interests amassed more than 30,000 acres of undeveloped land—Baker-Crow alone had 6,400 acres—mostly financed with short-term floating rate bank debt. In addition, the Crow interests had over 3,000 apartment units completed or underway, some of which were plagued by cost overruns caused by inflation and were not being rented easily. There were more than 5,000 apartment units that were operated at a loss, and 350 or so unsold condominiums in Atlanta. Interest charges alone came to $12 million a year, and were rising as rates ratcheted upward.

Baker-Crow had made important commitments in Texas and Oklahoma, where Willard Baker was gobbling up real estate as though interest rates were a minor consideration. In the meantime Crow, Pope & Land had expanded aggressively in Georgia and Florida. Both companies had been involved in cost overruns and had accumulated many unmarketable properties. In addition, they continued to operate in the old wheeler-dealer tradition. For example, Crow, Pope & Land was involved in the creation of the thousand-room Atlanta Hilton Hotel and a 300,000-square-foot office building project in that city, on which they began construction before the long-term financing was in place.

Baker-Crow and Crow, Pope & Land had large unrealized ("paper") profits on their land holdings, and this encouraged Crow to consider rising interest charges a necessary and worthwhile risk. Indeed, he embarked on new ventures in land accumulation.

In this period Crow became involved with farmland. Together with partners he had purchased a 43,500-acre tract in Louisiana called Angelina Plantation, which was really little more than trees and swamps. His intention was to clear the land and convert it to farming. It seemed to make good sense at a time when the price of such properties was rising. The subsequent bust was yet another headache for the Crow holdings.

Starting in late 1973 Crow received calls from both Baker-Crow and Crow, Pope & Land asking for infusions of cash. Despite these pleas and Glaze's more frantic warnings, Crow continued to accumulate debt. He began construction on Park Central, an office building complex in North Dallas scheduled to go up on 300 acres of prime land that cost more than $20 million, as usual financed through short-term, floating-rate loans. The interest on this sum came to over $2 million a year, and only a handful of the buildings were on line and generating income. Added to this were problems at the New Hebrides, Europe, and some farmland deals. Lincoln Properties was having the same kind of difficulties as Baker-Crow, with over $100 million in land debts. In other words, Crow was in trouble in almost all his operations except warehouses. What came to be known at the company as "The Crunch" had begun.

Crow finally had to concede that Glaze had been correct, and that as a result of his activities he was in deep financial difficulty. Glaze had the good grace not to remind Crow of earlier warnings, but rather tried to provide some semblance of liquidity to operations. Yet Crow's problems grew, and could not be kept from the partners.

Each partner learned about the Crunch individually. Ned Spieker's experience was typical. Crow's practice was to have the partners dispatch income to Dallas, which would service the debts from a central account.

One day Spieker telephoned Dallas to speak with the accounting department. He had remitted funds to the office, but his debts had not been paid. "Guys, you haven't paid the bills this month. I mean they just aren't paid. And I've sent you the money down," he said. When no satisfactory explanation was forthcoming, Spieker knew there was trouble. Glaze telephoned Spieker soon after and said, "Ned, the pool is closed." Spieker was puzzled. "What do you mean, Bob?" "Well, the pool account, the major pool account, has no money in it." Distressed, Spieker rejoined that he had recently sent some $2 million to Dallas. He was told that the money was needed in other areas. "I immediately recognized," Spieker reminisced, "that obviously there was a problem. Because the money was just going into one bucket and whoever could drain it first got it." Like the others, he expected that Glaze would play a major role in rectifying matters. Spieker did not realize at that time the seriousness of the situation.

Williams had no inkling about the problems until he asked Glaze for funds to cover a million-dollar shortfall in the Makro project being constructed in Brazil. It didn't seem a matter of great consequence, and so Williams was surprised when Glaze turned him down, saying, "Of course not. Every project has to stand on its own and be financed on its own." Glaze didn't say much more; Williams was relatively new to the firm, and Glaze figured that was all he had to know. As time wore on and the complications increased, Williams learned more about the problems and was included in the deliberations, and would play a commanding role in the bailout.

Crow would later remark that Williams was one of the most organized men he had ever met, and as had been the case with Glaze, the attorney complemented Crow's talents nicely. Crow needed someone like Williams to bring order out of the hodgepodge of operations he had put together. Glaze couldn't do it on his own, even had he possessed the stamina for the task. For several years he had wanted to retire, and remained on with the understanding that once the rescue was completed he would step down.

That Glaze and Williams were on the scene was fortunate for Crow, who by then had come to realize that while he possessed the talents to create an enterprise in good times, he was deficient in those required in the mid-1970s. This was another of those crucial moments in his life. While many successful businesspeople have a knack for knowing when and how to advance, few possess the self-assurance to concede the time had come to leave the scene to others better equipped to deal with the new set of problems and possibilities. Such was Crow's situation at that time. That he would play a key role in the workout was both important and obvious. After all, he was the central figure in the matters that had to be addressed. Crow left the strategies and tactics to others, however, and willingly accepted a subordinate role.

Glaze and Williams had a monumental task before them. By early 1974 Crow had entered into partnerships in warehouses, office buildings, shopping centers, and mobile home parks in the United States and several other countries. By itself Baker-Crow had 34 partnerships and 11 corporations of its own, mostly in land and apartment development. Crow, Pope & Land held apartments and condominiums under 21 partnerships and 10 corporations. Lincoln Property had 28,000 apartments, 3.6 million square feet of commercial space apportioned among 160 partnerships and 47 corporations, and 3 hotels. The Dallas Market Center was split between 2 partnerships (both exclusively Crow family interests), and the Park Central project among 4 partnerships. Then there were the Crow farms and ranches—12 partnerships there. Only the Trammell Crow Distribution Corporation, which was involved in public warehousing, was wholly owned by the Crow family.

In all there were 605 separate partnerships and 131 separate corporations, plus another 150 "in-house" partners and numerous other institutional or individual investment partners.

In most of these associations Crow was a minority partner. In some the partnerships were split between the working partner and one of the Crow children, this being considered part of their inheritance. As a result Crow had great personal debts for projects he did not control; any liquidation of the assets would have to be done with the full support and assent of his partners. These holdings were of considerable value, but could not be easily sold in that poor market environment.

Crow was then personally responsible for $151 million in borrowings and contingently liable for another $433 million. With these big liabilities he could under the worst of circumstances have been financially ruined.

The major task was stemming the negative cash flow, which toward the end of 1974 came to around $25 million a year. This money was due not to one or even several lenders. Each deal had its own partners and lenders, and there were hundreds of interests to satisfy, from small local commercial banks and mortgage bankers to the largest insurance companies in America.

The first step in the rescue operation was to inform Crow's major partners of the extent of his difficulties. This was done in a dramatic fashion. In early January, 1975, Crow called a partners' meeting in Dallas, which was attended by all the senior people, about 20 in all. Crow, Glaze, Williams, and Hope Hamilton arranged for the participants to be seated in a semicircle. The mood was decidedly somber. Glaze broke the news, which only verified the suspicions and knowledge some already had, that Baker-Crow and other holdings had serious cash flow troubles. Glaze put the matter directly. "Look, Trammell is having these problems," he said. "He helped you get started. He's your roots. Now he needs this group of friends and partners to sell some buildings and lend their share of the proceeds to him so that he can get through this period." Glaze's plan was to gather together properties that would be sold to the Equitable Life Insurance Co. and other investors at prices to be negotiated, and then use the proceeds to liquidate debt.

Most of the partners reacted to the news of Crow's crisis with surprise. Although some had had hunches for almost a year that all was not going well, none had any idea just how serious the situation had become. "The numbers were a surprise," recalled Kresko. "I was also surprised by the fact that we had people in the company who would have done some of the things that had been done." He went on to note that his own commercial operations were in good shape, and not even knowing some of Crow's partners in other sections of the country, he had supposed they functioned in the same generally conservative way he did.

You know, I assumed that everybody ran their operation with sound business practices in mind. And that indeed was the fatal mistake, to have assumed that. There were bad deals all over the country, where people had gone and overextended themselves. And I think they did so for one of two reasons. They might really have believed Trammell was capable of writing out checks for untold millions of dollars, directed by their aggressiveness, or they simply didn't give a damn. They might have thought they were on a crap shoot. They hadn't been worth anything two or three years earlier, and they were prepared to risk all they had amassed on a roll of the dice.

While Kresko and the others considered the matter, Glaze handed out copies of a list of all those present. Next to each name was a dollar figure that

Two leaders of the Trammell Crow companies in the 1980s: Don Williams and Joel Peterson.

Trammell Crow at his desk in the Stemmons Tower. Note the open office environment. Bob Glaze confers with a colleague in the background.

A company outing: rafting on the Chickamauga River, Georgia (1983).

A break in the action during a tennis match with former President Gerald Ford (1975).

Meeting with then Vice President George Bush (1987).

Trammell Crow in front of Bryan Tower (1980).

Loews Anatole Hotel in Dallas (1979): "A Village Within A City," and Crow's personal favorite project.

Gus Dubinsky, pictured outside the Brussels International Trade Mart.

The Infomart in Dallas (1984). Some recent Trammell Crow developments.

Texas Commerce Bank Building, Las Colinas TX.

Lighton Plaza, Kansas City KS.

Miami International Mall Promenade, Miami FL.

Hamilton Lakes, Itasca IL.

Jefferson Court, Washington DC.

Minnetonka Corporate Center, Minneapolis MN.

each was to raise through the sale or refinancing of properties. He then went around the room, saying, "Kresko, we need you to raise $2 million in your area, and Spieker, you need to raise $1.5 million." Now these were not modest amounts, and everyone knew what Glaze was suggesting. He was asking everyone in the room to undertake a self-inflicted partial liquidation by selling their most valuable and profitable properties. The partners would keep 25 percent of the proceeds to pay capital gains taxes, and would advance Crow the rest on his personal I.O.U.

Crow was to market $25 million of his own property. This was painful for a man who always believed that simply holding on to good properties in bad times and good would be extremely profitable. Now Crow would have to dispose of many of his prize holdings, even though he was convinced that in two years or so he would be kicking himself for having done it. And indeed, all those properties are worth much more today than they fetched then. However, he saved some properties, most notably the Market Center, which had a large positive cash flow throughout this period. Had all else gone sour, Crow would have that jewel still.

When Glaze came to his own name on the list, he announced that he would sell $6 million of his own properties. Later on, Crow's secretary, Barbara Collins, recalled, "It was a shock to all of us, but the general attitude was, 'Of course we'll go along with whatever has to be done.' " Gary Shafer, by then one of the senior partners, said, "I felt like I was a participant in a Super Bowl game. These were the people I wanted to be with, and there just wasn't any question in my mind that we would rise to the challenge. Whatever we needed to do, we did." Another later told a reporter that "I was 43 and unemployed when Crow brought me into the business. We're no kindergarteners. I'm willing to throw in my chips with Trammell." Bart Brown, the partner in charge of operations in Austin, San Antonio, and Oklahoma, was among the first to approach the shaken Crow after the meeting. "Trammell," he said, "everything I have is yours if you need it."

This self-sacrificing spirit was precisely what was required. The partners were being asked to endanger their financial futures on the possibility that their leaders would be able to somehow salvage a portion of their holdings. Under the worst of circumstances they would have relinquished properties worth millions of dollars and had nothing left but Crow's I.O.U.s. In sum they were risking their financial well-being in order to rescue other parts of the Crow empire with which they had no direct or even indirect interest.

They acted out of loyalty and gratitude, but also from self-interest. It would have damaged the credibility of every parcel of property with the Crow imprint had they done otherwise and the entire structure had toppled. By then one of the group's most important assets was the Crow name itself. Perhaps the partners hadn't given the matter much thought before, but now they did. For years they had prospered by being identified as Crow partners. Would they be able to do as well without that designation? This question too was going through their minds that day.

All the partners but one came through, though afterward several conceded they had hesitated. Williams and Glaze went through the room, asking, "Are you on board?" "How about you?" "Can we count on you?" Everybody nodded approval or said yes—everybody except Gil Thomas, Crow's first, largest, and wealthiest partner. When his turn came, Thomas reportedly said, "I'll have to think about it." All in the room were shocked. To this day Crow can recall the moment vividly. He felt betrayed, angry, and numb.

Thomas's net worth—somewhere between $5 and $10 million—was mostly in warehouses he and Crow owned jointly. Soon after the meeting he notified Crow that he would not accept the bailout arrangement and that he wanted to liquidate their partnerships. His eventual refusal to go along with the program Bob Glaze had developed meant that dozens of these properties could not be sold or used to collateralize loans.

Thomas's lack of support crippled the Glaze program, and also indicated to Crow that the man whom he had befriended and had helped make a multimillionaire, who had worked with him for 20 years, was prepared to see him suffer or even fail without offering his assistance. To make matters worse, Thomas next indicated he considered Crow dishonorable because he had not offered all his own properties, including the Market Center, for the bailout. Thomas engaged lawyers to make certain Crow was not pledging some jointly owned properties for other debts. He placed a freeze on Crow's part of the assets, and in other ways showed his disdain for the rescue program. Williams tried to reason with him, but all that Thomas could say was that his warehouse operations weren't part of the problem, and he wouldn't be part of the solution.

Under the circumstances the partnership had to be ended as rapidly as possible; Thomas had to leave the organization. The unenviable task of making the arrangements fell to Hope Hamilton, and the atmosphere in the office in which they worked out the details alternated between icy and red hot; either they wouldn't talk to each other at all, or, when they had to communicate, did so in vehement exchanges. Finally they worked out the intricate details of dividing several hundred properties. The matter of signing the papers began at 7:00 on a Saturday morning and continued on until 3:00 in the afternoon. Then it was over. Thomas took his warehouses, and Crow sold to the Equitable the ones he received from the partnership in the continuing effort to cut back on obligations and become liquid.

Among other holdings, the partners were to sell off about half of their 45 million square feet of warehouse space along with interests in some of the prime office buildings, which would bring in nearly $70 million. But the sales weren't enough; additional funds would be required. Nothing more could be obtained from his in-house partners. Crow had sold off all that could fetch a price. Frank Carter offered to buy Crow out of Crow-Carter, and this was done. New sources of funds were sought from out-of-house partners, banking associates, and others Crow had worked with in the past. First National Bank in Dallas provided $35 million, for which Crow pledged his remaining warehouses. The Chase Manhattan and Lomas & Nettleton came up with a series of major loans.

Some of the creditors purchased buildings or interests in them; Metropolitan Life traded its half interest in 2001 Bryan Tower for Crow's half interest in Allen Center. All in all, the Crow interests sold some $100-million worth of properties, and Trammell regretted each one that went. The Embarcadero Center holding went to David Rockefeller for $1 million, and a like price was realized for Douglas Plaza in Irvine, California. Park Central, Crow's dream to remake Dallas, had only five buildings when it was sold to the Equitable for little more than the investment.

It was a troublesome time, during which Crow was plagued by doubts and fears. "It's frightening to think that you owe $20 million, for example, and that it may not be there when the notes come due, and that one slip could bring you down," he later recalled. Crow would awaken in the middle of the night and stay up for a few hours simply staring at the wall. Occasionally Margaret would awaken too and they would play backgammon for a while. Then Crow would go back to sleep, only to awake again, restless, an hour or so later.

Crow jogged every day, kept his weight under control, didn't drink, and tried his best to keep his spirits up. He later insisted that all through this period he was basically confident he would survive. What might have happened had he lacked this confidence? Crow wouldn't have persisted in his efforts, others would have sensed his insecurity, and all attempts would have ended in failure. "I did not think of myself as a Pollyanna; I was never deluded into underestimating obstacles," he once observed. "But it is worthwhile to reflect that Pollyanna did all right for herself."

Crow's waking hours were consumed with trying to work his way out of the difficulties, and in the process he discovered who his friends were and how strong were their ties to him. Take the situation at the First National Bank in Chicago. He had a major relationship with that institution. Crow went to see the chief loan officer, who reminded him that he hadn't been living up to one of the provisions of a loan. Crow was offended. "Look," Crow told him, "there is no room in this discourse for that kind of statement. If this is the kind of conversation we're going to have, we're going to end it right now. Before we do, I want to tell you something. This company is fighting to stay viable. We will be a better borrower from you and from all of the other creditors if we do something to keep ourselves strong. I want you to know that we're going to negotiate in our own behalf in everything we do in order to stay strong. We want a relationship with First National in Chicago. We want to be a good borrower to you. But I can tell you that nobody in this office has done anything that is dishonorable. There might be a misunderstanding, they might be negotiating a point hard, but we are not being dishonorable."

From the look on the officer's face Crow feared that his words and his tone of voice had pressed the banker a trifle too hard. But the loan officer said, "Well, Trammell, those are the words that we need to hear because we want to do business with honorable people, and there is no doubt in my mind that what you said is true."

Other creditors were dealt with in different ways. For example, Pogue was having problems at Lincoln Properties, owing money to scores of creditors, with two in particular having large sums due them that simply couldn't be raised. So he went to talk with them, laying it on the line. Pogue described his situation, spoke of the amounts owed others, and then told of his intention to pay the amounts due everyone else but those two. That would keep his credit intact—so long as the major creditors took no action or spoke of the matter. In effect, they would be permitting Pogue to roll over his loans, and when the misfortunes were surmounted, he would pay them off in full, and with interest. If, on the other hand, they pressed for payment immediately and openly, he would be forced to seek bankruptcy protection from the lenders, which would mean their payments would come in after a long period of time, and probably only a few cents on the dollar. Moreover, they would lose a valued customer and associate, and the entire matter wouldn't sit well with the real estate community in Dallas. Those two creditors had more to gain than lose by going along with Pogue. So they did. Lincoln survived, as did their relationship.

By this time rumors of Crow's problems were causing some concern among investors in Lomas & Nettleton, which was reflected in the price of its common shares. In its annual report for the year the company revealed that as of October 31, 1974, it had committed $118,499,000 to Crow and his partners. The figure declined to $79,732,000 by August 31, 1975. The company devoted seven pages of its annual report to its relations with Crow, which was so important to its health. CEO Jess Hay observed that Crow accounted for almost a fifth of all Lomas & Nettleton loans, and tried to assure investors and others that "unlike many in the industry, Trammell Crow has *both the resources and character to weather the pressures of the current depression in the real estate development industry and he will do so*" [italics added]. It was a remarkable document, reflecting not only a businesslike analysis of loan structure, but personal confidence in Crow as well, and Hay's public support of Crow was heartening.

> *But what the rumor mill consistently has failed to recognize is the fact that Trammell Crow is not only the country's largest developer of real estate; he also probably is the largest individual holder of good income-producing real properties in the United States. And, although it is true that Trammell Crow's liquidity has been severely strained by the pressures of the depression which currently afflicts the development industry, it also is true (assuming sanity among his creditors) that Trammell Crow has the requisite resources to restore liquidity to his operations.*

Crow was taking steps to do just that. In this period he sold off part of the land inventory by forming joint ventures with Gibraltar Savings & Loan in Houston and with Equitable that produced $2 million in cash and relieved him of $30 million in debt. A group of Kuwaiti businessmen with whom Crow was putting up a hotel complex in Atlanta helped by buying out his interest for $2 million to cover cost overruns.

It was a struggle, but that autumn Crow thought the situation was improving, and that they were going to make it. His friends were most encouraging and supportive. Jess Hay sent him a memo in September titled "General Comment on the Rumor Mill."

Trammell Crow has been engaged in real estate development since 1947. He has always been dynamic, aggressive, optimistic, and highly leveraged. These characteristics frequently have led to premature announcement of his imminent financial demise.

Thus, the rumors have revived from time to time:

Circa 1957: He'll never survive the market complex folly.

Circa 1960: He'll run out of cash. He'll be broke in a year.

Circa 1969: Trammell Crow and Lincoln Property have experienced cost overruns of $25 million. Nobody in real estate can survive that much. It's just a matter of time.

Circa 1973–75: Real estate has gone to hell in a handbasket. Trammell Crow is the biggest name in real estate. Therefore, Trammell Crow is en route to financial collapse.

Forbes magazine disagreed with this analysis, and on October 15, 1975, ran an article titled "Crow Eats Crow," which catalogued his problems. While generally accurate, the tone was depressing. The situation was actually not as bad as the *Forbes* reporter indicated. The skein of friends and associates started to become edgy, yet none of them wavered. In fact, several went out of their way to indicate their continuing support. Mortgage broker Glenn Justice, whom Crow had known almost from the beginning of his career, took out a full-page advertisement in the *Dallas Times Herald.* The ad featured a picture of Crow, and across the top were the words, "Columbus Discovered America. Guys Like Trammell Crow Help Build It."

The only actual transaction which developed into an immediate and substantial threat came from a real estate investment trust. Crow had a mortgage of $4 or $5 million on some land that was past due, and the money simply wasn't there. His attorney signed off, saying that nothing could be done in the matter. Crow was despondent. The foreclosure was set for the following Monday, after which creditors would swamp him with demands.

On Friday afternoon another attorney, Bob Middleton, came up with a clever solution. The state legislature was then sitting, and Middleton recalled a Texas statute that grants automatic stays of execution in all matters to companies or individuals represented by attorneys who are members of the legislature while the legislature is in session. He arranged for Crow's case to be taken over by a Dallas attorney who was also a member of the legislature. The necessary papers were delivered to the courthouse first thing Monday morning, and the foreclosure was blocked with only a few hours to spare. By the time the legislature adjourned several weeks later Crow had accumulated sufficient funds to repay the loan, did so, and thus saved both the property and his reputation.

Ten years later Crow sold that property for 10 times the amount of the loan. He often wonders just how much Trammell Crow Company would be worth today if he had been able to hold on to the other properties he was forced to liquidate. It would have been billions of dollars.

The crisis seemed unending. That summer some officials at First National Bank in Dallas began to doubt Crow's ability to come through intact, and were troubled about their substantial loans to his operations. One day, while attempting to renegotiate a loan with Rawles Fulgham, the bank's president, Crow was hurt to hear him raise his voice in anger, shouting, "You borrowed the money. You agreed to repay. Now don't you think you should do so?" It was clear he would not get the extension; Fulgham, who was about to retire, obviously wanted to leave his successor, Elvis Mason, a clean slate. First National was Crow's major banker. He had run out of borrowing capacity and had no one else to turn to. But he managed to get a delay, and together with Williams tried to figure out his next step.

Crow, Glaze, and Williams took the problems to the next partners' meeting that October, at which time they put out an agenda with the innocuous headings of "Lender Relations" and "Business Postures." Then Crow arose and told his associates just how grim the situation had become. He proposed taking all his remaining assets and applying them to outstanding loans, which would take care of all debts save those of Lincoln Properties. In effect, Crow would become a sort of lightning rod, drawing the heat and protecting the others. "I think I should stay on everything, and get everyone else off," he concluded. "What I end up with is of no concern. The partners, the company, will live. And I will come back again."

Later on, when the crisis was all over, one of his partners told Crow that this moment was the nadir for him emotionally. All that year he had thought they would be able to survive. "That was the first time I felt despair," he said. Hope Hamilton even thought that several of the partners had contemplated suicide. But all rejected Crow's proposal. Williams spoke for all of them: "I don't think the partners are safe until you are safe."

Now the partners realized that there was a real danger of bankruptcy. Given the general precarious economic situation, if Crow had gone belly up the reaction from lenders and the government might have been even more drastic than what had happened during the Penn Central debacle of 1970, when the credit markets came within a hair of panic. Indeed, through this period the Federal Reserve maintained a credit watch on the Crow interests, prepared to step in if disaster threatened. Had Crow failed, the consequences to the nation's economy would have been severe.

Fully cognizant of Crow's difficulties and knowing they would need some outside assistance, in 1975 Williams conducted a series of discussions with Kenneth Leventhal, head of a Los Angeles accounting firm that specialized in salvaging indebted real estate companies, a subsection of the industry that was then in great demand. At the time Leventhal was assisting Pogue with his liquidity problems at Lincoln Properties, and he had concluded that it was virtually impossible to consider Lincoln without taking the Trammell Crow interests into account as well. Pogue then approached Crow and suggested they employ Leventhal in both campaigns, and Crow readily agreed to do so. At that stage of the rescue, Crow was prepared to listen to almost anyone.

That initial meeting was memorable. It was at Turtle Creek, where Lincoln was domiciled. On the way there Williams ran a stop light on Lemmon Avenue and his car was broadsided. Deep in thought, he hadn't been watching the road. Fortunately, only his car was damaged.

Indeed, Crow and his colleagues were all troubled before the meeting, expecting the worst. But Leventhal provided a healthy antidote to their fears. He was a smart, wise-cracking accountant who had been in such situations before and had emerged unscathed. Leventhal had recently come to the aid of another Dallas company, Redman Industries, and he had performed wonders at Boston's Cabot, Cabot & Forbes. Both companies had been in situations somewhat similar to Crow's. They were still around, he said, and there was no reason why Crow shouldn't prevail as well.

In the course of the meeting Leventhal convinced the Crow contingent that their problems not only weren't unique, but that they could be resolved. He argued that the book value of the assets was almost irrelevant.

The germane figures were debt and cash flow, and the under-the-hammer values of the properties, opposed to their supposed worth. That was where he started with earlier clients, and where he would begin at Crow. The next step was to come to a judgment. "We attempt to get our arms around an entity and ascertain: (1) whether it is viable, and (2), can it be reorganized and saved." In both instances Leventhal's responses were in the affirmative.

From the beginning Leventhal inspired trust and affection. He obviously understood the positions of creditors and debtors. He was tough-minded, able to outline the situation exactly and to offer solutions. At another time, under different circumstances, Crow would have tried to enlist him for his operations; he remarked that Leventhal was and is one of the best salesmen he had ever known.

One day soon after the meeting Leventhal informed Crow that some of his holdings verged on what he called "substantial embryonic judicial insolvency of considerable magnitude," which was his way of saying they were bordering on bankruptcy. These holdings could file for a Chapter X, XI, or XII bankruptcy, he mused, as Crow sat back, horrified at such talk. And then to Crow's immense relief Leventhal dismissed this as unrealistic. He observed that any bankruptcy proceedings would take the better part of a decade to complete. The legal fees alone would eat up whatever capital came through intact. The senior creditors would see only a fraction of their money, the juniors would get nothing at all, and the partners would be destitute.

Leventhal talked of the possibility of bringing all Crow's creditors together in one room to discuss the situation. He planned to take a hard line with them. He had used this approach with the other workout cases, but it couldn't be done at Trammell Crow, Leventhal said with a straight face, because there were so many creditors he would have to rent a football stadium for the meeting. Initially he thought they might be apprised of the situation and presented with an ultimatum: Either go along with the painful restructuring of debt and obligations or accept the bankruptcy. By then Leventhal had learned that the Crow interests had more than 700 business entities in 30 American cities and 8 foreign countries, with liabilities approximating $2 billion. It would be the most massive workout in history, and he counted on the creditors' self-interest to win the day for his proposal. In effect he would throw the keys to the properties on the table and dare the creditors to foreclose.

Crow, Williams, and the others could not accept this harsh suggestion. In addition to the financial stakes there was the matter of trust and honor. A large number of those creditors had been doing business with Crow for years, and many were personal friends. To treat them so cavalierly would have violated their relationships and deeply offended Crow's personal sense of integrity and decency. Leventhal appreciated the qualms, and generated an alternate plan. He now suggested that together with Williams and Crow he meet groups of creditors and explain the situation. Even this would shock the creditors. The more sophisticated of them would know that the various Crow interests had borrowed from many others. What would be their reactions on learning that there were hundreds of others beside themselves?

Leventhal called this plan a "nonjudicial workout," and convinced Crow that it could be done. If *every one* of his creditors agreed to it, they would be safe. But if a single creditor insisted on payment, they would be obliged to seek protection for that unit under the bankruptcy laws, and when word of this leaked to the press, the other parts of the Crow empire could come tumbling down.

Seeing that Williams and Crow thought the task formidable, Leventhal whipped out a chart Peterson had prepared for him detailing the Crow holdings and relationships (the same chart found in the "Introduction"). It looked like a convention of spiders—lines going every which way, extraordinarily complicated. It could well have served as a Trammell Crow coat of arms. It showed them how interrelated the Crow ventures had become and illustrated the central theme that all were bound together. Leventhal intended to use an augmented version of it in presentations before the lenders. "When the creditors realize just how complicated it all is," he said, "they will realize that cooperation is preferable to litigation."

This marked the beginning of a new stage in Crow's attempts to rectify the situation. At first he and his partners had attempted to satisfy creditors' demands by selling properties. Generally speaking, those creditors who complained the loudest and made the most threats were paid first. There was a constant hemorrhaging in Dallas, with funds obtained by selling worthwhile properties being used to pay creditors who held loans on less desirable ones. They reacted without any real strategy or plan. Now Leventhal provided one. There would be no more property sales. Instead, together with Crow and Williams, he would attempt to work out settlements and delays.

In the months ahead Leventhal hammered away at convincing creditors of the need to work with Crow on his salvage operations. He would explain the situation to a lender, and then produce that chart. Leventhal would say, "Look at this. You don't know where you are, and I'm not sure I can find you there. Now I've asked you to come here because I want to give you a haircut [meaning a tough deal]. Don't bother to look at what's on the other guy's plate. Just look at your own."

Leventhal would then go on to show how Crow and his partners could work things out, attempting to assure the lender he would receive full payment if he cooperated, but that everything would fall apart for *him* if he insisted on payments at that time. In effect, Leventhal was saying that the lender had no choice but to give Crow time and even additional credit. He possessed the ability to smile and quip without diminishing the seriousness of the situation.

The method was simple enough. Each of the major partners would deal with the creditors with whom he was most closely associated. The partner came to the meeting with what was called the "Gray Book," a 45-page, legal-sized folder that Leventhal and his staff had prepared, in which were presented arguments and evidence to convince readers of his ability to survive. The partner displayed the organization of his company, and showed them that tangled web Leventhal had developed. Next he would show them the values in the company, answer questions, and then obtain agreement.

Williams and Leventhal worked as a team, and did so most effectively. They were also getting help from some of the partners who, unable to develop new business, were recruited for staff work. It was a difficult switch, for most of these people had become accustomed to running their own operations and working deals. Now they had to function as middle-management personnel, and almost all of them bridled at the change. For example, instead of acquiring land and arranging for the erection of warehouses, Shutt was sent to Europe to try to raise capital for the entire Crow interest. "I felt like the middle guy on a toboggan," he recalled. "I couldn't steer and I couldn't slow it down. I had spent too many years making all of the decisions on what we're going to build, when we're going to build it, to make the transition."

When "the Crunch" was over, Shutt would leave the company to work on his own in oil and natural gas, not in real estate development. But, like the others, he bent his efforts at that time to the tasks of raising money and stilling creditors' fears. Meanwhile Glaze coordinated efforts from the Dallas offices and acted as troubleshooter. When he was unable to do all this and attend to his always overflowing financial tasks as well, Crow hired

Jarmon Bass, formerly an Arthur Andersen partner, to become chief financial officer.

Often the lenders tried to get a better deal for themselves than Crow was offering the others, but Leventhal and Williams insisted that each be treated the same, that that was the only way it would work. Patiently and persistently they made the case for restructuring the debt, making clear the advantages. To indicate their earnestness in a tangible and humorous way, Williams ordered some two-color kneepads similar to those worn by basketball players, with the company's name imprinted on them. He and Leventhal were willing to get down on their knees to pray for deliverance if that was required. Toward the end of the presentation Leventhal would walk around and say, "Hey, guys, this is it. We want to cut a deal with you." Then they would depart, leaving it for the lawyers to work out the details.

One of the most arduous encounters was in Atlanta, where Crow was having difficulties with an increasingly overwrought A. J. Land. At one point Land referred to Crow as no more than an antecedent general unsecured creditor, meaning that Crow's rights were subordinated to those of the others. This marked the beginning of the end of their relationship. Then, without Crow's consent or knowledge, Land had one of the Crow, Pope & Land condominium projects file bankruptcy. The news spread apprehension among the partners, who feared the creditors would suspect that Leventhal's plan had disintegrated and would stampede to the courtrooms. Crow and Williams had to reassure everyone and let them know they were not in a liquidation mode. They knew that everything depended on atmosphere and confidence, and something like this could cause irreparable harm. Once the word was out that they hadn't changed the game plan, things settled down. But the undermining of the rescue efforts spelled the end of Crow, Pope & Land. They agreed to separate their assets, with Crow taking all the apartment projects, while Pope and Land received cash and warehouses.

Now Crow had to locate someone to take charge of his residential holdings, which were in the process of being reorganized as Brentwood Properties, and what remained of Baker-Crow. Both operations were dispirited. Trainees either left to seek other employment or were dismissed due to lack of business. In 1973 there had been more than 70 of them; a year later only a dozen or so survived, most of them junior people. Tom Teague, who arrived in 1973 as a summer intern and took a full-time job in 1974, with less than a year's experience assumed responsibility for 2,000 units, about twice what the normal load should have been. Personnel was stretched thin, and morale was low.

Given all the bloodletting, there were so few trainees left in the residential area (and none on the senior level) that Crow was obliged to look outside the organization for a new leader. He found Terry Golden, a native Dallasite who was raised a few blocks from the Crow home and even knew some of the Crow children, but who had never met Crow until then.

Golden had impressive credentials, but until comparatively recently in his career he had had only a peripheral interest in real estate. After earning an undergraduate degree at Notre Dame, he went to Massachusetts Institute of Technology for an M.A. in nuclear engineering. Then, after a spell at General Electric, Golden returned to Boston for a Harvard M.B.A. Upon graduating he took a post at Babcock & Wilcox, initially as chief financial officer for the company's nuclear fuel plant, where he was responsible for production control. Golden's next assignment was in Europe, where Babcock & Wilcox was engaged in a joint venture with Brown Bovari manufacturing and marketing nuclear power plants, and once again he was in the financial sector. Golden left Babcock & Wilcox to go to Puerto Rico as financial officer for a troubled subsidiary of the Sea Pines Company, a land development company. He received the offer from Ron Terwilliger, a classmate at Harvard who then was that firm's chief financial officer. Within a year and a half Golden was president and part owner of the subsidiary, and he had decided to switch careers and become a land developer. He also made investments in Columbus Realty, which was Golden's initial exposure to real estate, this bringing him into contact with officials at Chase Manhattan.

By then Golden had enough of overseas living, and wanted to return to Dallas. Asking his contacts at Chase if they knew of a suitable position, he was directed to Crow. After a few meetings, Crow decided Golden had the proper outlook and attitude to take charge of one of the residential holdings he intended to rebuild, Lexington Properties. By the time Golden arrived in Dallas, however, Crow had sold Lexington, which meant there was apparently nothing for Golden to do at headquarters. After several meetings with Crow and Williams, however, it was decided to turn the entire residential problem over to him. Golden's experience in financial matters and in working his way out of difficulties at the Puerto Rico operation would be useful in attempts to do the same for the residential properties.

Brentwood and other residential companies would have greater need for one with Golden's background than a person who was primarily a builder.

While Golden struggled to familiarize himself with Crow's operations, Williams, Leventhal, and the others continued their travels from one creditor to another attempting to work out mutually satisfactory arrangements to defer payments. All involved had to take a lot of abuse. When Williams was making his presentation to a savings and loan in Waco, the elderly president grew increasingly agitated and walked over to Williams and shook his fist in his face. "Sonny, just pay the interest! That's no hill for a climber!" Williams tried to explain that Crow could not pay, but the man simply grew more livid with every word. In the end he complied, but Williams had to work for that one. For the most part he retained his composure, but there were times he had to let off steam. For example, every day a loan officer from a New York bank called to find out how matters were progressing, and on one occasion flatly accused him of lying. Williams slammed down the telephone, called the officer's superior, and warned that if he wasn't taken off the case, the Crow interests would let the matter go to court. The officer was switched.

Most creditors were cooperative. First National of Dallas was one of the first to sign on. Elvis Mason had replaced Rawles Fulgham by then, and he just came in, took off his jacket, rolled up his sleeves, and said, "O.K. Let's see how this can be done. We're going to help you through this." And that is what they did. What came out of this meeting was a plan Crow's negotiators could take to the other lenders and say, "Look, First National in Dallas has agreed to this method *subject to substantial compliance by all the lenders*." Each time they convinced one to agree, the presenter would add, "[So and so] is in agreement also, *subject to substantial compliance by all the lenders*."

In general, steady progress was made. Yet there were always lenders who were suspicious. Crow later learned that Denny Wallace, one of First National's vice-presidents, came across some of them at a mortgage bankers convention. As expected, the main topic of conversation was whether or not Trammell Crow would come through intact. Wallace assured them he would, because Crow had enough money to meet all his obligations. When asked how he knew this, Wallace replied, "Because I am going to give it to him."

The Chase Manhattan was a good example of how matters went. Crow owed it over $100 million, and naturally they were deeply concerned. By then it seemed that everyone knew the Federal Reserve was tracking the rescue attempt, so contacts at the Chase were fully informed. They never had to call Crow; someone at the Dallas headquarters telephoned them regularly and frequently. Of all the major creditors, Leventhal recalled, those at Chase were the most understanding and cooperative, perhaps because they had gone through the procedure several times in the recent past.

Of the more than 250 creditors, all but 3 cooperated. Jess Hay helped with one of them. Crow was past due on the principal on a note for $2.5 million, and the lender's attorney telephoned to threaten legal action. Had the news come out, others would have followed, setting off a rush of panic that could bring everything down. Crow spoke of the matter with Hay, who called the attorney and offered to buy the note at a $750,000 discount. He would give him a day to decide, and the alternative would be to come to Texas and get involved in a long, drawn-out court battle. The creditor took up the offer, Lomas & Nettleton purchased and extended the note, and Crow was able to wipe out some debt. Hay continued to provide assistance during the next two years, and this was a crucial element in Crow's recovery. It also was good business for Lomas & Nettleton, for the company profited greatly by backing him in this period.

At times Crow seemed to be standing still, and there were moments when he thought it would never end, that he would spend the rest of his life trying to straighten things out.

This is not to suggest that all Crow did in this period was to contract operations. Far from it. Not only did he and the commercial partners continue construction programs begun before the crisis developed, but they didn't let anyone go; everyone remained on the payroll. Continuing to put up new buildings for the market when the need justified it proved to be a very smart strategy. It was to his good fortune and a commentary on the wisdom and nerve of the better lenders that those who really understood and appreciated Crow's circumstances continued to advance money for new projects. Moreover, he was having little trouble attracting equity investors. Ehrenkranz placed client funds in Crow undertakings in Chicago, Minneapolis, San Francisco, Dallas, Houston, San Antonio, Atlanta, and Denver. In the process Ehrenkranz, Ehrenkranz & Schultz became the largest equity investor in the commercial side of Crow's enterprise.

Toward the end of 1976 the pressures started to ease. For one thing the partners had paid off or recycled most of the debt, and for another interest rates had started to fall. Everyone involved was realizing that they would survive after all. The ebbing of the crisis showed in Crow's greater liquidity. By 1977, for example, his debt to the Chase was down to $20 million, and by 1978, $10 million. The largest nonjudicial reorganization of an American real estate development company ever attempted was just about over. Friends and rivals had always said Crow thought big, and he had to agree. But this wasn't what he had in mind.

The financial crisis had crystalized Crow's position: Never again would he permit himself to get into such a bind, but at the same time he was not going to permit the experience to make him timid, unwilling to accept the kinds of risks that success in the business requires. True enough, a different kind of company emerged out of the Crunch, with different people and a new kind of organization. The business hadn't died in 1975 and 1976, but it could be said to have been reborn in 1977.

On the surface the Crow interests hadn't changed much. There were still hundreds of partnerships and companies, each legally separate from the others. The period of adversity and forced cooperation had brought them all together emotionally as well as financially. Under the direction of Glaze, Williams, and Leventhal all those boats, large and small, had been linked together. They still might operate to an extent on their own, but not completely, for ties had been created that not only would persist, but be enhanced.

It was a different kind of company; in fact, it was on its way to becoming a *company*. This entity would require full-time leadership, strategic planning, tables of organization—the panoply of business that Trammell Crow had always found so distasteful, and for which he had little talent. In his 60s he was still vigorous, active, and hungry—but for deals, not management opportunities.

EIGHT

The Trammell Crow Company

The essence of the Trammell Crow Company is a paradox. We have found a way to organize entrepreneurs. By nature entrepreneurs wheel and deal. They do things on their own, not in concert. You're not supposed to be able to organize them, but we have been able to do that. As our partners have prospered, so the company has prospered. If I've been able to contribute anything, it might be a skill at building consensus, keeping all these really good partners doing well on their own, and at the same time keeping a sense of common goals. That's what Trammell Crow did for me and for so many others, by being what he is.

—DON WILLIAMS, 1985

Early in 1976 Bob Glaze prepared to retire and pursue his private concerns. He originally had intended to do so in 1974, when he was 55 years old, but he put off retirement until he was reasonably certain the Crow interests were back on track after the Crunch. Now he started to empty his desk.

Crow reacted to the news characteristically by attempting to dissuade him. There was no reason Glaze could not have a desk at the office, he argued. In that way the two of them could at least see one another on a regular basis. Jarmon Bass already was functioning as chief financial officer, so those responsibilities no longer need concern Glaze. All Crow wanted was for him to be there for companionship and consultation. Glaze firmly rejected this proposal, believing that in a few days "Trammell would have filled my basket without my ever having any time for myself." So he left, and this key departure symbolized the fact that the entire Crow enterprise was about to undergo major changes.

While the Crunch had been a sobering blow, initially at least Crow had no intention of permitting it to drastically modify the ways he conducted business. True, there would be more fiscal prudence than before at headquarters and in the field; the wild and free-spending days of Crow, Pope & Land and Baker-Crow would never return. Nonetheless, Crow still balked at the thought of creating controls that might check his activities, or anything that would get into the way of making deals.

As far as Crow was concerned, the old ways that worked so well in the past could be modified somewhat to serve in the future. All that was required to set matters right would be a few alterations mandated by the Crunch and post-Crunch circumstances, plus the replacement of Bob Glaze by Bass or some other person capable of handling details after Crow set down the essential elements of his deals.

Now 63 years old and as vigorous as ever, Crow was eager to get on with development—additions to the Market Center, hotels, office buildings, trade marts, and the like. Warehouses too, of course, but he no longer was as interested in these. Some of the regional partners who had proven themselves during the Crunch constituted the most experienced and capable crew in the industry. They could take care of their bailiwicks, Crow reasoned, while he roamed the country, seeking targets of opportunity.

But Crow's vision of the future could not be. The Crow interests could never return to the way they had been managed prior to the Crunch, and one of Don Williams's more important tasks was to convince Crow of this.

By 1976 it had become obvious to most within the Crow constellation that Williams was assuming additional responsibilities. Kresko recalled the precise moment he knew that Williams was about to move into a more prominent position at the Crow interests. It was August 29, when he received the partnership papers on a Kansas City deal from Dallas. As usual, he received a 30-percent share, and the local partner was given 20 percent. Ordinarily Crow would have received 50 percent, but in this instance the entire amount was assigned to Williams. "Now I don't know if at that point Don had ever been in Kansas City, but it didn't matter," said Kresko. "Crow had the right to assign a share to anyone he wanted to, and he had often done so in the past. But never before had it been the entire 50 percent. Clearly Williams was special."

Crow would have preferred Williams to have functioned as his new Bob Glaze, which is to say take care of all details while Crow negotiated deals. Williams rejected this role, if only because that niche had been tailor-made for the older man, and no one could truly replace him. Williams wanted to convince Crow that the old ways simply wouldn't do, that the new business environment called for a different kind of structure, and that Crow himself would have to assume a new role at the company. Crow would have to adjust himself to a new administrative style. As Shafer put it, Crow's way of conducting business was, "You make a deal, put it on the shelf, go make another deal, put it on the shelf and just keep going." Given the size of the business and the altered economic environment, such an approach was no longer appropriate.

One of Williams's initial tasks in the post-Crunch months was to tell Crow that these modifications in attitude and techniques were necessary, that in order to preserve the quintessence of the Crow spirit in the new business environment changes would be required. He put it succinctly several years later:

> *When I saw, during the Crunch, just how vulnerable the whole organization was, it seemed obvious to me that something had to be done to change it from a group of talented partners operating on an* ad hoc *basis with the help and blessing of a head entrepreneur, into an undertaking of entrepreneurial partners working together as a business. Where there had been one man with a series of joint venture relationships with a number of people, there needed to be a group of cooperating partners organized for the long haul.*

The more Crow considered Williams's arguments in favor of change, the more plausible his appraisal seemed. Therefore, on Williams's initiative as well as Crow's, Williams was being prepared for an entirely new kind of position.

From both experience and observation Williams had developed a keen understanding of the altered situation in the country and world. More specifically, the real estate industry was quite different from what it had been when Crow and his partners had put together their various holdings. The time of low and relatively stable interest rates had ended, at least for the time being. So had the era of exuberant growth that Crow typified as much as anyone else. His approach to commercial real estate development—that it was an enterprise in which nothing was impossible if only one were determined and willing to work tirelessly—had to be tempered by a more sophisticated and less ambitious perspective. Arrangements to put up warehouses on speculation, without visible lessees, made sense in the 1950s and 1960s, and perhaps would serve in the 1970s as well, but such speculation was now foolhardy for the wide range of commercial and residential properties. Williams was convinced that the company's thinking along these lines would have to change.

The hedging of risks and the generation of liquidity, hardly Crow characteristics in the past, would become more important at the company. There was a new sense of limitations in Dallas, even while everyone was planning a much larger company in the future. Crow always appeared eager to recreate the world, to alter circumstances to make them fit his vision of what they should be. Williams, on the other hand, perceived a business climate in which the Crow interests would have to adjust to shifting circumstances, to respond to events more often, perhaps, than causing them to happen. He recognized diversity in conditions and opportunities in widely differing regions and nations. The sensitivity and subtleties of approach he had developed while an attorney, which had been honed by Crow experiences in Europe and Asia as well as in the United States, would serve the community of interests well.

Williams recognized that the Crow operations themselves had changed considerably, not only in kind but also in scope. From the dozen or so employees Crow had in the late 1960s the business had expanded to nearly 200 by 1977, and almost 90 percent of these had been with the company less than six years. By then Crow's commercial businesses had more than 25 offices, up from the 7 at the beginning of the decade. In this same period the number of partnership arrangements had risen from approximately 100 to more than 700.

The experience of the Crunch had persuaded Williams that in the future greater coordination and communication among the partners would be needed. The imagination and intuition that Crow had demonstrated in conceiving the Market Center, office complexes, and more functional warehouses had to be supplemented and informed by a more elaborate and reliable system of gathering, monitoring, and utilizing information.

In Williams's view, the revamped Trammell Crow management needed an administrative structure that could institutionalize cooperation without enforced conformity, encourage individual initiative without running the risks of irresponsibility, and retain an abiding local marketing orientation while guarding against stifling bureaucratization. If the company was to thrive, the essence of its entrepreneurial qualities must be preserved while order was imposed on the partners' activities. In short, for the Trammell Crow operations to continue to be as strong and vital as they had been prior to the Crunch, they would have to come to terms with change. Williams knew the general direction he had to take. Working out specifics, refining his objections, and winning approval from the partners would be his major concerns in the years that followed.

The Crunch had become to the collective psyche of the Crow operations what the Great Depression had been to America as a whole, namely a horrific experience to be avoided at almost all costs. To continue the analogy, to some of the more troubled partners Crow was akin to Herbert Hoover, the person in charge at a time of grave difficulties. At this point the comparison breaks down. Whereas Hoover was later castigated, Crow still retained the admiration, affection, and respect of the partners even though many of them were wary of his activities. Still, the man John Stemmons once called "our wild little brother" could not be allowed to have his operations come to the brink of collapse ever again.

The required changes were begun, not from Dallas but from the field. These changes bore the seeds of friction. For one thing, Williams moved Spieker's accounting operations from Dallas and Spieker then conducted financial operations out of his San Francisco office. He and several other key partners were pulling into their own shells and attempting to segregate their operations more fully from Dallas, while Williams was striving to achieve greater coordination of efforts. They were tugging in different directions, and this tension would continue into the 1980s. As the Trammell Crow interests expanded and were transformed, the underlying independence/coordination issue remained. It is there today.

As Williams saw it, what had made the Crow operations unique was that the heads of the local offices were full-fledged partners, encouraged to move aggressively in pursuit of *their* goals. If good fortune prevailed, their moves rebounded to the benefit of the aggregate company as well. This was the perception of the partners too, but for most it went even beyond that. Each Trammell Crow partner, in each city and region, believed he *was* the company, and that the particulars of the local situation were all important. There was no overarching view, no time horizon beyond the next few years. Few were aware of what might be accomplished by a larger and more complex entity in which they were a part but not the central focus, and for that matter, there was slight concern with such questions. So the partners saw little need for the kind of strategic planning Williams felt necessary if the Crow interests were to flourish in the new environment.

In the pre-Crunch period Crow had said to the partners, collectively and individually, "You know what works best in your own territory. Go out and develop real estate, and count on me to come up with the money as you need it." The result was the creation of a vast confederation of real estate interests. Given Crow's nature, there had been no one in the organization to make certain the partners in the field always knew how their activities would affect the Trammell Crow operations as a whole. In retrospect it became obvious that if all the partners had known of the problems as they developed, they might have conducted their activities differently, and perhaps avoided much of the pain.

The partners appreciated that what had saved the company was Crow's acuity in selecting associates. He had assembled a set of partners who in the Crunch turned out to be capable not only of negotiating a successful rescue operation, but of cooperating with one another in the enterprise. What had been a collection of interests was compressed by the Crunch into an entity in which members were aware of other parts of the enterprise. Men who formerly had been concerned only with, say, warehouses in one part of the country, and who didn't even know the names of some of the other partners, now kept in contact with them on a more or less regular basis. They also collaborated on intricate maneuvers, learning to rely on one another, and started to develop a degree of collegiality.

Yet even without this searing and bonding experience, the Crow interests would have had to come up with some means of cooperating with one another. The market itself was changing, and Crow would have had to realize this and adjust to the new business climate.

For example, this was a period when real estate companies were realizing that opportunities for corporate mobility were growing. In different parts of the industry, Coldwell Banker and Cushman and Wakefield could provide services for companies and businesspeople moving from one market or part of the country to another through coordination of branch offices. Such information existed only in a rudimentary form at Crow in the mid-1970s. If a partner in one city knew a client planned to move to another market, he might have communicated the information to a counterpart there, but as it was, little of this kind of communication took place. There had been no real incentive for the situation to change. Now there was.

As Crow came to appreciate Williams's vision of the company's future, he also realized that not only would he be unhappy in such a business environment, but that some of his qualities that enabled him to perform so spectacularly before the Crunch were perhaps inappropriate in the second half of the 1970s, and probably beyond. More than he craved success, fame, and power, Crow wished to perpetuate the enterprise, and then hand down to his successors a company that would outlast even those buildings he had erected. This would be his true legacy. "I wanted it to exist permanently," he said, "as a secure and strong member of my country's business community, and provide rewarding lifetime careers for those who worked for it." Crow came to understand that if this goal of continuity and renewal over the years was to be achieved, the overall administrative operation would have to be self-correcting and adaptable to changing circumstances. He came to a difficult and inescapable conclusion: His future utility to the operation would be as counselor and door-opener, not as chief executive and operating officer.

Both Williams and Crow recognized that Crow's ultimate value to the firm at this point was as an example to the young trainees certain to arrive in larger number over the next few years. Given the turnover in personnel and the expanded and diverse opportunities bound to be created, there had to be some way to infuse the newcomers with the Trammell Crow spirit of service, integrity, and decency. One could try to capture the spirit in manuals and disperse it at the meetings that now became more common at the firm, but manuals and meetings could not replace the man himself.

There also were major operational tasks ahead. The commercial side of the business would have to be restructured. The residential business, in shambles, would have to be recreated almost from the ground up. New divisions would be needed to capitalize on new opportunities. Crow's presence would be needed in all these areas. Williams hoped what would emerge would be a new Trammell Crow Company, a unique set of enterprises still in need of the constant presence of the namesake.

From the first, the operative concept was that it was necessary to organize and rationalize the business, particularly the commercial side of it. The new plan would accomplish both. The buzz word in Dallas and in the field then was "evergreen." The collection of commercial partners would be fused into a more permanent structure, not dependent on any single person.

These concepts percolated in Crow's mind in 1976 and into 1977, during which time he talked about them with partners and other trusted friends, as well as members of his family. Then, together with Williams, he began developing a plan to reorganize his holdings into three nebulous but at the same time distinct entities: the Trammell Crow Company (TCC) for commercial development, the Trammell Crow Residential Companies (TCRC) for apartments and residential lot development, and finally the various Crow family holdings, including the Dallas Market Center, warehouses, hotels and in time hospitals as well.

These were not companies in the traditional sense of the term. The old collection of deals was not to be disturbed, for even had Williams and Crow wanted to, they would have faced strong opposition. If anything, the experience of the Crunch had instilled in many partners a wariness about leadership from Dallas. The commercial partners had almost been dragged down by failures outside of their territories and even their lines of business. Determined this would never happen again, they intended to retain more of their capital at the home base rather than sending it on to the common pool at headquarters.

Partner sovereignty was bound to be a touchy issue in the years that followed. Yet the partners appreciated the new stress on reporting and information gathering, if only because had such instrumentalities been in place earlier, the Crunch might have been avoided. In Williams they perceived a leader who would insist upon such procedures, the kind for which Crow had little use or interest. While willing to accept additional direction and coordination from Dallas, they wanted more financial autonomy in the field within the overall philosophy of the firm. Even so, the field partners were unaccustomed to perusing long memos, considering strategic planning outside their own territories, and responding to inquiries and filing forms.

So the Crow operations were being pulled by competing forces, those calling for a more centralized operation and others hoping to maintain local control. Moreover, there were tensions among the three component parts of the enterprise. The partners at TCC, who dealt solely with commercial properties, were uneasy with anything in the residential field, and were decidedly negative toward those in that business who operated under the Trammell Crow label. This sentiment militated against further amalgamation at that time. Indeed, quite a few TCC partners would have been pleased to sever all ties with TCRC. Finally, the family interests were not truly separate, but united in the sense that the Crows had positions in all the deals at the company and TCRC. The family concern constituted a preserve that brooked no interference by the other partners, which created some problems. To resolve one of them, Williams urged Crow to separate himself from other ventures with which he was associated, often competitive with each other.

Even before enunciating his strategy, Williams found it necessary to create the beginnings of an overarching structure through which an agenda could be devised, and then debated, defined, and implemented.

The first step was the formation of a board of partners to direct the new enterprise, to be known as the Trammell Crow Management Board, made up of senior partners representing the various commercial regions. The initial board members were Williams, Crow, Shutt, Allan Hamilton, Spieker, Shafer, Kresko, Simmons, Brown, and Mark Myers. The board would formulate policies for the company and then attempt to win approval. It had limited powers, in that while it set down strategies, the individual, local real estate decisions were made by the local partners, who retained their regional power. Dallas might persuade, but it would have difficulty coercing.

In the board, however, the various Crow commercial concerns finally had a unified operating executive force. Yet all the members except Crow and Williams were still local or regional developers, each controlling his own territory like a fiefdom. Of course Crow continued on as a developer of other projects such as hotels. As might have been anticipated, Crow found committee work boring and frustrating, and after a few meetings he stopped coming.

The initial design was complete and ready to be placed in operation in June, 1977. Crow now gathered his commercial partners and informed them of what most already knew was to occur. Effective immediately, he was stepping down. He would continue on as a partner, working with the other senior partners, under the leadership of a managing partner who would also be in charge of commmercial development: Don Williams. Unlike Crow, Williams would work full-time at this task. In his time Crow had handled dozens of deals simultaneously while coordinating efforts in Dallas in what amounted to his spare time, often on the run. A major reason for his earlier difficulties was Crow's unwillingness to give up his deals in order to concentrate on the operations as a whole. Williams would not have this conflict: while willing to aid in deals, he did not consider them his primary job. He would have the power to fire partners, in effect transforming them into limited partners. Williams exercised this right, which triggered a series of buyouts. In turn, he could be removed as managing partner by a vote of the board.

Soon thereafter, to further unify the operation, the commercial company created Trammell Crow Partners, which would receive 10 percent of three out of every four deals, to be placed in a common pool that would be divided on a regular basis, several times a year. This meant, for example, that Tom Simmons in Houston would have some common ownerships and interests with Bob Kresko in St. Louis, and vice versa. It was an extremely difficult and delicate task, which however was fairly readily accepted and implemented.

The same kind of philosophy guided developments at the Trammell Crow Residential Company, where the tasks were even more delicate. Williams's counterpart at TCRC had to erect a company out of the shards of what remained from several failed enterprises and a few other holdings. The tasks at the family holdings were quite different, but there too changes in leadership and structure would be required. In addition, the matter of relationships of both the family and the residential holdings to TCC had to be ironed out, taking into account the sensitivities and concerns of all involved.

Each of the three components would have its own outside partners, objectives, and methods of operation. TCC would be more patterned than TCRC, while the Market Center and other Crow interests would be operated pretty much as they had been earlier, and would not become an integral part of the TCC.

For the moment Crow gave precedence to the creation of a viable structure at TCC. He went out of his way to emphasize his subordination to Williams. "Trammell placed his imprimatur on Don," Joel Peterson recollected, "and then didn't meddle." Now it was Williams's turn to lead the partners in his own way, which would not and could not be Crow's. Williams appreciated what this signified. "I'm not a leader like Trammell, a charismatic, original thinker," he asserted. "That's not my way. I find outstanding people going in the right direction and get behind them as a leader. Our goal in Dallas is to serve the field organizations, to help bring out the best in the local programs, and do everything we can to be supportive. Not to direct and control what they are doing. What I do is listen, and see that when someone has a good idea it gets shared."

As Crow had done before him and like most CEOs, Williams believed his primary task was to develop proficiency in his partners, to teach by example, and to seek out others who could also perform such tasks. Employing an analogy from baseball, Williams saw himself as a professional manager. "Part of my job is to keep the team strongly staffed, which also means seeing that those who can't perform up to standard aren't kept on as part of the team. We've got a tremendous amount of talent in this company," he said, "and my responsibility is to do my best to see that all that individual excellence gets a chance to fulfill itself as part of a championship team." Williams's manager analogy is fitting, since baseball is a sport in which players have to perform individually on some occasions, and on others function as parts of the club. This team/standout duality was what he hoped would develop at Crow.

Williams would not have to start from scratch. But the task of harmonizing the inherited bits and pieces was an imposing responsibility, for he would have to concoct structures that would release and channel energies without disturbing those elements that had made Trammell Crow's components so vigorous in the past. This meant Williams could not create a centralized corporation, even if that were his intention, which it was not. Rather, he would first try to create a sense of solidarity, and then see what could be done with the rest of the Crow interests. The next step was to develop a methodology and vehicle for planning, reporting, and maintaining contacts. This task was assigned to Peterson, who clearly was to play a significant role at the refurbished company.

Williams, filling the slot left vacant by Crow, needed someone to perform the duties of chief financial officer once executed by Glaze. Williams did not believe Bass could operate effectively in the kind of company he had in mind, for by then it had become apparent that Bass was uncomfortable in the free-wheeling world of developmental real estate. Williams asked Peterson to take the post, which he did.

The Williams-Peterson relationship would be different from the Crow-Glaze mode of operating. Williams did not require from Peterson the cautionary assistance Glaze had furnished Crow. Peterson's tasks would not include completing the details for Williams's deals, for the new CEO had no intention of operating as Crow had. Then too, Glaze and Peterson were quite different in outlook, interests, and training. The older man was an accountant who was comfortable with profit and loss statements, bookkeeping procedures, and the like. It was this background that enabled him to function as an anchor for Crow's vaunting ambitions. Peterson, on the other hand, was a Harvard M.B.A. with special interests in finance, unequipped to take Glaze's post, and in any case far more interested in strategic planning. So he wasn't to act as a Bob Glaze to Williams's Trammell Crow. Rather, the two men would operate in tandem. Williams provided strategic leadership, was the key figure in relations with field partners, and developed additional sources of capital. Peterson tended to focus on tactics, accounting and reporting standards, and deal analysis, along with the fashioning of infrastructure. Furthermore, Peterson was still quite young and relatively inexperienced. For several years he was involved in learning the ropes, coming along rapidly, until by the mid-1980s he had become one of the three most powerful people at the firm, and perhaps the most knowledgeable regarding all of its details.

Peterson and Williams had developed a good working arrangement, and both men realized they complemented one another. Williams was more intuitive, a born conciliator and organizer, tolerant of imperfections and diplomatic in pointing them out. Peterson was superb at details and problem solving and was something of a perfectionist. So while the working relationship of Williams and Peterson differed from the Crow-Glaze model, there were also points of similarity, especially in personal style. In general, Williams concentrated on external affairs, while Peterson concentrated on internal matters. While making crucial decisions about strategy and personnel, Williams also launched a wide-ranging revamping of company practices.

Peterson was charged with the creation of accounting, tax, financial, and administrative controls for what otherwise would remain a decentralized operation. In 1977 he embarked on a company-wide study to isolate weaknesses and offer suggestions to rectify them. Problems were many and diverse. Most of them could be traced to the lack of planning and extremely short time horizons in the past. For example, there was a marked lack of consistency in financial reporting. Some partners regularly provided Dallas with ample information, while others communicated only occasionally or incompletely. A number of partners could predict with some certainty when problems might arise; more could not or did not do so. Lines of authority and responsibility were unclear; too often tasks were left undone because those responsible thought someone else was accountable, while in other circumstances redundancy was the rule. There was a generalized lack of back office efficiency. Most important, the disorder fed into the atmosphere of chaos and crisis, generating a desire on the part of some partners to become more cautious and risk-averse, which if continued could result in a decline in business and a loss of opportunities.

The planning, reporting, and accountability problems would have to be addressed expeditiously and with vigor. There was to be a Management Review Plan, under the terms of which annual meetings were scheduled between Williams and Peterson and each of the partners for a line-by-line assessment of that partner's financial and real estate performance and plans. Throughout the year the two remained in close contact with the partners, gathering information, backing their operations with additional capital sources, deal analysis, and back office support. Order and discipline were important for the businesslike operation of the firm, evaluating actual quarterly financial results against budgeted projections, and then analyzing the variations.

Obviously the success of this type of planning depended not only on what the Management Board decided upon, or even on the development of a system of information generation and reporting data; it also required the willing participation of all involved, especially the local partners who did the developing, selling, and leasing. Independent thinking was built into the company's partnership structure, and that would remain; decentralization and entrepreneurial initiative had been key to the company's performance, and these elements too would be preserved. Now a greater measure of teamwork would be added to the mix.

Personnel was a sensitive problem. Over the years Crow had added management positions and partners somewhat haphazardly, with the result that there were redundancies in several areas and gaps in others. The Crunch had revealed the weaknesses of several commercial partners. Crow tended to be uncritical of those in whom he placed trust, which meant that several people who were relatively unproductive managed to rise in the organization. They would have to be eased out or moved into different, less-demanding areas—and a way would have to be found to ensure that in the future only high-performance people reached the partnership level.

Williams initiated a thoroughgoing housecleaning. Superfluous employees and offices were eliminated, and weak operating partners replaced. Territorial and job responsibilities were altered and defined in specific language (in many cases for the first time), as were reporting channels and company compensation policies. The partnership agreements, central files, and information systems were organized and extended to ensure swift access to required documents and financial information. Williams developed contingency plans for all manner of human eventualities, including death, retirement, withdrawal, and conflict within the partnerships.

In the 1960s Crow had created an employee profit-sharing trust. Now the profit sharing was expanded so that nonpartner employees would have a greater stake in the business. From then on an interest in one out of every four projects would go to the trust. In this way the concept of partnership was enlarged to include virtually everyone at the firm.

All the compensation changes were set down in black and white. What formerly had been casual understandings were now carefully drafted and articulated statements. In most cases there were no substantial differences involved, in practice. But the actions denoted as much as anything else the contrasts in the Crow and Williams styles—that of the free-wheeling real estate wildcatter had been superceded by one that might be expected from an attorney.

Many of those involved expressed doubts about the wisdom of documenting the partnership agreements—Crow himself questioned whether policies should be committed to paper in such a fashion. Might not codification of procedures result in paralysis of will? Williams insisted it be done. The old Trammell Crow business had been a reflection, elaboration, and extension of the founder; the new company had to be able to stand on its own, dependent on no single individual. As it became independent it required what amounted to a constitution and by-laws.

During the late 1970s there emerged a group of senior partners and regional organizations to go with the review and budgeting mechanisms and limited controls from Dallas. As Peterson put it writing from the vantage point of 1985:

> *Ownership patterns were established and other generally agreed-upon policies were embraced. Many of the partners that "grew up in the organization" during this period are in positions of leadership today. Other partners who could or would not meet production or other leadership standards either resigned or were asked to leave. Partner meetings took on aspects of training as well as camaraderie. The Management Board emerged as the governing body of the organization.*

For the first time, the Crow interests demonstrated a strong concern with posterity, indicating they were planning not only for years or even a decade, but for what these new senior partners hoped would outlast their tenure at the firm. Williams demonstrated his intention to retain the old but to regularize it in his approach toward the recruitment of new people.

Note has been taken of Crow's inclination to base his staffing decisions more on instinct than on personnel analysis. This was changing somewhat in the late 1960s, when the stream of newcomers arrived from the M.B.A. programs. Williams continued and accelerated the movement toward a more formalized recruitment and training effort. Assuming successful interviews with the local partner, newcomers in the commercial field were put to work as leasing agents at one of the company's regional offices. The salary remained low; by the late 1970s it had risen to $18,000 a year plus commissions, with the hope of an eventual partnership. Recent experience indicated that they could expect to make $50,000 in the second year and around $80,000 or more in the third. Those who demonstrated genuine ability were next given the chance to expand into site selection, finance, construction management, and, in time, building design. The objective was to create a consummate development partner, a generalist who could initiate and then manage a large operation, open and staff a new office, and undertake a project of any complexity or scale. Partners would also have to hone their leadership and interpersonal abilities.

After three or four years, the trainees reached what was in effect an "up or out" situation. There were no hard and fast rules then by which partnership status was conferred, but in general the agent considered for partnership had to successfully meet several criteria. Most important, he or she had to have demonstrated exceptional abilities in marketing—lease negotiations, tenant services, and the generation of new business through resourcefulness and creative and innovative thinking. In the construction area the candidate should have shown ability to design products for clients and potential renters, to work with architects and contractors, and to bring in projects on budget and on time. These abilities presupposed a good working knowledge of real estate law, accounting, finance, and construction management. Potential partners were expected to be hard-working, flexible, and to have consistently high levels of both energy and imagination. TCC wanted real estate specialists, but at the same time expected its candidates to have other interests as well—a concern for political and economic issues affecting the industry and business, for example. It has been seen that Crow had an abiding passion for art and architecture, and other senior associates had their own nonbusiness-related interests. In sum, the criteria for partnership were demanding, and as time passed they became even more so.

If, in conjunction with other partners and the Managing Board, the local partner decided the trial period had been successful, trainees would be elevated to "project partners" receiving a small equity interest in the projects to which they were assigned. The next step upward—and there was no specific timetable for it—was for them to be offered a divisional partnership, which implied that they were now expected, in part at least, to generate their own projects. Assuming all went well and openings existed, the divisional partner might be elevated to a regional partnership, which would mean supervisory duties over divisional partners and an equity interest in their undertakings. Finally, they might aspire to become national partners, who were charged with operating TCC. The goal was to create a national company. Ever since Trammell Crow first burst on the scene the company was viewed as a Dallas developer. One of the objectives pursued in the 1980s was to provide the firm with a reputation more in line with reality.

Initially most of the partners did their own hiring, with mixed results. By the early 1980s, as competition for talent intensified, Dallas took on added responsibilities in this sphere. As in other large companies, headquarters dispatched partners to leading graduate schools of business, seeking top talent who would then be referred to field partners for interviews or internships. Recruitment was difficult; many of the M.B.A.s graduated with family responsibilities and student loan debts. Offers of employment at twice the salary tendered by Trammell Crow, along with more status than was inferred by the title "leasing agent," must have tempted many who eventually opted to come to Crow in the expectation of being made a partner if everything worked out well.

Given the nature of the real estate business, quite a few recruits must have considered that their ultimate objective was not a regional or national partnership, but a business of their own. As has been seen, it was not at all unusual for Trammell Crow partners to go off on their own, and there was more of this to come. In the 1960s (and in the 1970s and 1980s as well) Crow believed that if handled properly, the separation could be beneficial to all concerned. The departing partner would have made his or her contribution, and the departure opened opportunities for the younger partners.

New partners and old alike faced a more structured method of withdrawing funds from the firm, which may have hastened some departures. Prior to the reorganization partners wishing to withdraw some of their capital had to talk it over with a senior partner, who in turn brought the matter to Crow, who either approved or disapproved, usually without specific guidelines. Now the partners knew they had to maintain a certain level of liquidity, and that withdrawal would not be permitted if it meant a decline under the cut point. This was just one more example of how Williams and Peterson strove to impose guidelines and rules on an operation that previously had little restraint or regulation.

The issue of compensation was also debated and discussed and, as noted earlier, documented. What evolved was a tiered system of rewards. By the early 1980s Williams and Peterson would receive 7½ percent each for all projects. The Crow family interests would receive a varying amount, depending on circumstances, which averaged 15 percent. The national partners participated at 10 percent, and another 10 percent went into the profit-sharing fund. The regional and local partners, who actually created the project, arranged for its financing, oversaw the construction and got the rest of the deal, with a substantial portion of that further parceled out to those within their organizations. The Management Board oversaw the fair division of shares. In case of shortfalls the Crow family or other senior partners would provide needed capital, especially for newer partners. As Crow later put it: "The paternalism here is very complete. We are father."

As noted, Terry Golden became managing partner for the residential companies subsidiary. He was involved in development as well. However, his major tasks at TCRC were in the areas of salvage and rebuilding. This done, he would not proceed along the same lines Williams was taking at TCC, but rather would develop methods more suitable for the residential business. The new TCRC did not intend to engage in the kind of long-term, comprehensive strategic planning familiar at most large enterprises, aspects of which might be useful at commercial real estate companies as well. The business on the whole simply did not lend itself to such structure, being deal-oriented rather than product- and enterprise-centered. Indeed, crude attempts at long-term planning were what had led to so many problems at TCRC, when moves were made to stockpile land for future needs. Each regional partner instead was to present annually a thorough budget and account for projected cash sources. Attempts would be made to retire working capital debt and so establish TCRC's financial strength and credibility. In addition, specific methods were established for separation through conflict, retirement, and death of major partners from TCC.

Under the circumstances one might have anticipated a measure of apprehension and even suspicion regarding Management Board intentions. From the earliest, all experienced what was to become a characteristic of the new TCRC—dynamic tensions between the individual and local ambitions of the partners and the designs of Williams, Peterson, and others on the Management Board. Properly cultivated and astutely managed, these tensions could be enormously creative and invigorating, calling forth the best efforts of all involved. If handled clumsily and unimaginatively, however, the opposite could have come about, and the relationships could become adversarial and frustrating, resulting in stagnation and separation. During the following years, the company would know both.

As indicated, in the beginning TCRC had only the slightest direct relations with TCC, whose partners considered Residential the weakest part of the enterprise, its most expendable business. Crow did what he could to squelch such sentiments. When a Management Board meeting occasionally turned into a TCRC-bashing session, Crow would become irate, defending the Residential partners as "good people," guaranteeing that he would support them, asserting they would do more for TCC than TCC ever did for them. This defense would stop the talk, but the feelings remained. As far as the people at TCC were concerned, TCRC still was not really part of the family. Healing this breach, creating confidence in TCRC capabilities in the minds of TCC partners, was yet another concern of the Management Board. In part this would be taken care of by the passage of time, and clear indications that TCRC was being prudently and profitably shaped by Terry Golden and Tom Teague.

Initially, however, the situation at TCRC was neither prudent or profitable, but difficult and complicated, requiring drastic surgery. A beginning was made in 1976 with the sale of a 50-percent interest in Brentwood Properties (the new name for Baker-Crow's multi-unit dwellings and what remained of Crow, Pope & Land), which had been transformed into a property management company, to Stanley Fimberg, a Beverly Hills lawyer turned real estate syndicator. The partial divestiture relieved some of the financial pressures, but others remained.

Of paramount importance was the matter of morale and staffing. The entire experience with residential properties had left sour tastes in the mouths of most of the Crow partners. Throughout the Crunch all realized that the warehouses, hotels, and office buildings were for the most part solvent, while the major losses came from Crow, Pope & Land and Baker-Crow. Viable, worthwhile commercial properties had been sold, often at distress prices, to repay loans on the residential side of the business. Additional holdings would have to go. Furthermore, many of the residential partners had left the firm, and there were few left to participate in any effort at reconstruction.

Terry Golden had little to work with at TCRC. The companies he had been allocated were small, scattered, and troubled, the remnants of an ambitious attempt to fashion some of the nation's largest housing complexes. In virtually every case Golden had to restructure their finances. Workouts with lenders continued long after such accommodations had been completed for the commercial ventures.

Golden developed a method for dealing with problem holdings. He and others would survey the properties to determine their condition and prospects, their value, and the status of the partner. They would analyze management and the financial structure. Where necessary they would change management, and when that wasn't possible would disassociate from the partner. Golden would take whatever steps necessary to work his way out of the situation and develop a strong management structure.

He had to deal with a grab-bag of properties. There were those multi-dwelling units along with some undeveloped real estate left over from Crow, Pope & Land and Baker-Crow buying binges. TCRC had relations with scores of outside partners, and of course now with Fimberg, who at the time was content to remain in a fairly passive role. One major objective would be to raise sufficient funds from sales to become financially viable. Another was to demonstrate that Crow had a future in the residential area. Finally, Golden had to find ways to convince the commercial partners that the residential side of the business would not drag them down as it had earlier. If he succeeded, then a venturesome person might conclude that advancement could be more rapid at TCRC than in any other part of the Crow company.

This element could be seen at the beginning, when in February, 1976, the largest segment of TCRC, Brentwood Properties was turned over to Tom Teague. Teague had barely progressed from the ranks of interns, and now he was being handed one of the most important positions at TCRC. His initial task was to supervise and wind down the remaining construction work and to manage the assets that were left, trying to improve their operating performance. Finally, Teague had to be prepared to sell certain assets should problems develop at other parts of TCRC.

The assignment was at the same time complicated and simplified by developments at Lincoln Properties, on its way back but still plagued by creditors. Difficulties emerged in Crow's relationship with Lincoln and Residential. For one thing, Lincoln had developed into one of the nation's largest residential real estate developers, and as such was in direct competition with units of TCRC in some markets. Pogue and Crow had no problems with one another, but it was otherwise with the regional partners, who quite properly complained about conflicts of interest. Then, too, Pogue was at the point where he wanted to go it alone, which was both natural and to have been anticipated. This was a prelude to the divorce of Crow and Pogue—in a business sense, not personally.

The Crow and Pogue separation seemed in some respects to be a replay of Crow's divorce from the Stemmonses, though of course there was a major difference in that Pogue was going to buy Crow out of Lincoln. There was never a problem regarding the division of holdings. Crow acquired properties worth an estimated $200 million as his share in the company, and these went into Brentwood. They were to be shaped up, and some of them prepared for resale to raise additional funds so as to preserve solvency. Golden orchestrated the overall restructuring, after Teague made his evaluation, securing as many of those assets for Crow as he could under the circumstances.

In 1978, TCRC organized Crow Development, really that part of the discontinued operations that owned and developed raw land, the segment that had gotten the entity into so much trouble. The new unit was to be headed by Michael Crow, one of Trammell's nephews, who had been with the company for several years. As with Brentwood, priority was placed on solvency, but Development also fashioned a dissimilar, more conservative approach to land management than that of Baker-Crow or Crow, Pope & Land. Gone were ambitious plans for large-scale inventories. Instead the company would acquire smaller tracts to develop immediately, often with outside partners. Development would have the land rezoned for residential construction, if necessary. Then the land would be cleared, streets and utilities installed, and plots offered to builders. This was the extent of the operation—entirely horizontal, nothing vertical. Baker-Crow had done this too, among its other operations, but usually intending to sell to the upscale market. Crow Development focused on the moderate-price range, where builders might take a hundred sites rather than two or three.

By 1979 Golden and Teague had succeeded to the point of being able to consider expansion again, which meant bringing in some new development people. One of the first was Charles Holbrook, who was to be given responsibilities to develop multi-dwelling properties in Texas, Oklahoma, and Louisiana. This required the creation of a new entity headed by Holbrook, known as the Chasewood Company. It consisted initially of those Baker-Crow properties not placed under the Brentwood umbrella. Due to the recession of 1980 and 1981 Chasewood had a slow start, but when the economy improved so did business, and the firm soon was involved in construction throughout the central part of the nation.

The next new TCRC development partner was Ron Terwilliger, who had worked with Golden at the Sea Pines Company. A Naval Academy graduate with a Harvard M.B.A., Terwilliger demonstrated strong leadership ability. He was to operate out of Atlanta in what was named Crow-Terwilliger. The initial assets consisted of Crow's restructured Atlanta holdings, essentially of multi-family units inherited from Crow, Pope & Land. Though concentrating on the Southeast, Crow-Terwilliger's territory was the eastern third of the country. Initially Terwilliger converted rental property into condominiums, but within a year he had started developing multi-dwelling complexes, mostly garden apartments, as well.

In 1982 Golden asked Teague to supervise the Houston operations and organize another company, which became Crow Western, whose area encompassed the western third of the United States. With this the essential structure was completed. There was Golden in Dallas, Terwilliger in Atlanta, and Teague in charge of the West, each with his own territory, each taking on young partners.

TCRC was far from being a unified, coherent entity, and indeed was much less structured than TCC. But the same new fiscal conservatism that informed operations at Commercial was present at Development. For example, Crow-Terwilliger would guarantee individual partner's construction debts, in this way realizing significant savings. After a while, Development used corporate guarantees to limit liabilities in the long-term financing to very short periods of time. The company stressed liability management, which paid off in less exposure to the kinds of problems that resulted in the Crunch.

Golden became the link between the company and the Crow family interests. He was named chairman of Trammell Crow Distribution Corporation (TCDC), which it will be recalled was in the warehouse and trucking businesses and, unlike the other Crow interests and companies, was not a partnership but a traditional corporation. Employees received salaries and bonuses, and profit sharing was not the rule.

As noted, TCDC originated in 1969, prior to the Crunch, when Crow acquired several small public warehouse companies in Dallas and Salt Lake City. This was a logical expansion of the Crow warehouse interests. Just as TCC continued to erect and lease warehouses, so TCDC would own such facilities and leased them to customers, most of them in the "Fortune 500," on a less-than-long-term basis.

TCDC ran into financial difficulties during the Crunch, and in 1977 was still in need of reconstruction. Golden lacked the expertise and certainly the time to direct the company, so together with Crow he recruited Emory Wellman from the Anderson-Clayton Company, who in February, 1978, became president and chief operating officer. Both Golden and Wellman were given equity positions in TCDC.

Under Wellman's direction TCDC was turned around. The company expanded throughout the East and Southwest, the goal being to become a national entity. Wellman attempted to meet all of his customers' warehousing needs, to the point of erecting structures wherever they might be needed, and building to specifications. There was both a natural synergy and potential conflict with TCC. On occasion TCDC contracted construction to Crow Company, and Crow Company might recommend TCDC to clients requiring warehouse services. But they were in the same business, which meant that in some locales they were in competition. This anomaly was one of several that demanded attention during the 1980s.

One of Wellman's more significant innovations was the creation of a trucking business allied to the warehouse interests, capable of moving clients' goods from one locale to another. This was followed by entry into freight brokerage, which meant the firm could take a truckload of goods from, say, Dallas to Oklahoma City, and have a load there to carry back to Dallas on the return trip. The drayage was done on a commission basis, and after paying for such items as fuel, tolls, and salaries, was almost pure profit. As a result of such developments, and aided by an expanding economy, TCDC within a few years grew to be the fourth-largest entity in its field. It remained a family concern, separate from the other Crow interests.

In addition to his work at TCRC, TCDC, and the Management Board, Golden joined the board of Market Center, thus increasing his exposure to the Crow concerns. These remained pretty much as before. He would serve as a liaison among the various Crow interests until his departure for government service in 1984. Golden became Assistant Secretary of the Treasury, and then headed the General Services Administration.

Crow continued to develop his properties through the family operation, handling them as he had prior to the Crunch, which is to say, in a more free-wheeling fashion than was becoming the rule at TCC. It was as though Crow had handed the commercial business over to Williams with his blessing, and then turned around and continued to function as he wished through the family holdings.

Initially Bill Cooper continued to manage the Dallas Market Center, which would remain a family holding, but Crow's son, Trammell S. Crow, and daughter, Lucy Crow Billingsley, became vice-presidents. They took over when Cooper retired in 1983.

Even while Crow was facing the need for massive liquidations, it seemed obvious that new marts would be called for. The next additions came in the mid-1980s. Trammell S. Crow was monitoring several industries to see if any of them merited a mart of its own. One of these was menswear. There was no single special-purpose mart in the nation specifically devoted to this large area, although a New York City office building had been converted to showrooms. There were quite a few menswear showrooms in the Apparel Mart, with pressure from tenants who wanted to expand and new customers who wanted to enter. Lucy Crow Billingsley decided that if a Menswear Mart could be added, it would start out with those established tenants, who would vacate space in the Apparel Mart for others wishing it. This was done, and the Menswear Mart, a 400,000-square-foot facility housing more than 2,000 lines of men's apparel, opened for business in 1984.

Another, more glamorous area was computers. "At the time there were many computer shows, but none at which the entire industry had representation," Crow recalled. "There would be shows for mini-computers, large mainframes, software, communications systems, and so forth. Well, we were trying to make sense of that." Crow learned of plans to erect a computer mart in Boston, and he decided to act quickly. In 1982 he announced his intentions to erect a high-tech showplace. He took on as architect Martin Growald, who was known for having designed several unusual and intriguing structures. Growald and Crow decided to model the new building, which was to be known as the Infomart, after the Crystal Palace Exposition of 1851, which is to say in a Victorian style. Just as the Crystal Palace celebrated invention in its time, so would the Infomart in the late 20th century. Trammell S. Crow believed that the coldness and hard science of the computer would best be balanced by something warm and familiar. So the interior of the Infomart has wooden floors, potted plants, and other elements of Victoriana.

Two years after the initial announcement, the Crows celebrated the opening of the million-square-foot information industry showplace, the largest and most complete permanent display of computers and the most heavily trafficked exposition in the world, with over 350,000 equipment buyers a year going through its doors. Even so, it was not an instant financial success, going through some hard times until the concept was accepted in the late 1980s. Other additions to Market Center were to come. And Crow will probably continue to plan marts until he runs out of land or industries they can serve.

The same forces that impelled Crow to create new marts also mandated the development of a major new hotel. During the late 1970s the growth of the Homefurnishings Mart's business was such that attendees had to be lodged as far away as Fort Worth and Denton. Projected growth of business for all the marts indicated that the room shortage would continue. So the need for another facility to house customers was becoming increasingly evident.

At the time Dallas lacked a truly distinguished modern hotel like the 1,000-room giants that helped draw conventions to New York, Chicago, Los Angeles, New Orleans, and other major hubs. If Dallas were to be competitive it too would have to have them. The Hilton interests, which had made plans to close a downtown facility, were already considering a site near the Market Center, and Crow wanted to beat them to it.

Even without these considerations Crow might have gone ahead anyway. Hope Hamilton, who knew his habits as well as anyone, once remarked that "[w]henever he had pressing money problems, that was the day he was going to go out and find something quite big to develop. . . . He cannot exist very long without developing something." Crow conceded as much. "I'm hooked. I just can hardly let an opportunity escape to create something great, when I also believe it will be economically viable." Not only would the hotel be a major project that would enhance the jewel in his empire, but now he could turn aside from those tasks of reconstruction that so pained him. "The prospect of the pleasure of a positive, fun, planning, constructing involvement saved me," he later wrote. "We'd had three years in the trenches, fighting, struggling, defending, asking help and favors, selling, reaching for cash wherever we could find it. And my soul cried for the spirit which could only come from a positive project. The Anatole was that."

Building hotels was not a novel experience for Crow. He had been associated for a while with La Quinta Motor Inns, and of course had been involved with John Portman on the Hyatt in Atlanta and with Pope's ventures in Hong Kong and the New Hebrides. After these had come another Hyatt in San Francisco and a Hilton in Atlanta. The passion Crow once demonstrated for warehouses and exhibition halls by then had been transferred to hotels.

Crow and Bill Cooper and their wives traveled together to China on a business trip in 1976. While boarding the plane carrying them from Canton to Beijing, Cooper mentioned an idea of his about how to finance a hotel on an 18-acre site owned by John Stemmons across from the Market Center. Stemmons had an agreement with Marriott, which earlier had built a small Market Center hotel, under which that company had the right of first refusal on all land owned by Industrial Properties. Marriott had passed on its option, so Cooper thought Crow might be able to purchase the property. Crow and Cooper discussed the matter all during the four hours to Beijing. As they were about to land, Crow asked Cooper to see Stemmons when they returned to Dallas and find out if the land was available.

Cooper approached George Shafer at Industrial Properties and learned Stemmons was willing to strike a deal. Crow and Stemmons had little trouble arranging for a long-term lease. And so the way was clear to go ahead with the planning. As noted, under Cooper's direction the Market Center had done very well through the Crunch, and because of this Crow was able to use it as collateral for the loans needed to construct the Anatole.

Crow thought of the Anatole as a monument. While the Market Center was certainly impressive, it grew gradually, one building after another, and did not present the opportunities for planning that the Anatole did. Crow came to consider the project the capstone of his career. That it came at a time when Crow was turning over command at TCC to Don Williams may have been important; in a way, planning and executing the Anatole was a form of re-creation for him.

Despite his flamboyance and sometimes seeming disregard for finances, Crow had always designed and built with an eye toward costs. The atria may have seemed a waste of commercial space, but Crow utilized them in the belief that the higher rents the buildings commanded more than justified the expense. Now, in his new incarnation, and just this once, he would push aside all considerations of expense. The Anatole would be built in a lavish style that in an earlier period Crow would have deemed improvident. Not only was it to be one of the world's great hotels, but in it he would recapitulate a lifetime of concern with art and architecture. Crow felt strongly that such a project should come after, and not coincident with, formative learning experiences. "Sooner or later, most developers, and many other people, aspire to plan, build, and own a hotel," he later reminisced. "Fortunately for the industry, and also for the aspirants, few do so." The reasons are the risks involved in such projects. Crow reflected:

> *My principal observation is the construction of a new hotel of this type is not for youngsters. It takes a cool head and a long purse. A hotel has most of the dangerous aspects of real estate and of an operating business, and few of the good aspects. Hotels must be promoted, and their use sold. Most group business is sold years in advance, and if you are opening a new hotel, you've not had any such bookings. And even if you had, most meeting planners would not book a hotel for future meetings until they could see it in place. Then, besides sales, there is management, operations, food and beverages, etc. It takes about one employee per room to properly operate an active medium-to-large hotel. That's a lot of children to feed.*

In other words, the risks involved in such grand projects are immense, but the need for great hotels obliges developers to assume the risks. "Of the various types of real estate projects, few if any radiate as much support and value to adjoining properties and projects as will a quality hotel. Consequently, almost all large or mixed-use commercial projects strive to have one."

Crow turned the architectural assignment over to Overton Shelmire of Beran & Shelmire, who had been the architect for several of the Trade Mart projects. Within weeks Shelmire had a preliminary design and estimate. In the original plan Shelmire envisaged a large, 10-story-high rectangle with a very large atrium in the middle, bifurcated by a "land bridge" of guest rooms. In effect, it would look like a domino tile.

Crow was dissatisfied with that design proposal. So Shelmire embarked on trips around the country, to inspect recently constructed hotels and talk with architectural scholars and consultants. Several other plans were presented, most retaining the concept of having two atria connected in one way or another. In the end Crow himself came up with the final design. There was to be a 13-story-high square atrium adjoining a much larger 9-story rectangular atrium. This approach opened new vistas for Crow, allowing him to build one atrium as a formal "room" and the other as a more casual one.

The initial atrium plans featured a good deal of water and trees. Shelmire thought Atrium I would have a reflecting pool that would take approximately half of the space, while Atrium II would be heavily wooded. Crow thought there was too much water and foliage; they would get in the way of the people. He wanted the guests to feel as though they were indoors in a formal, palatial setting with natural light coming from the skylights. The final solution featured 50 large ficus trees, hundreds of plants, and a fountain and pool in each atrium.

There would be a lounge in the middle of the first atrium and a restaurant centered in the second. Other restaurants and shops, meeting rooms, and halls would line the walls. In all, it would be a 900-room city under a roof, which bore some resemblance to the squares found in many European and Latin American cities, major shopping malls, and urban downtowns. If the concept could be realized, on busy evenings, when conventions were booked and the city entertained many visitors, the Anatole would have hundreds of people walking across the atria, dining and drinking, heading toward one or another meeting or convention. Later on the Anatole would promote itself as "A Village Within A City," and with some justification. It became a place for Dallasites to meet for an evening on the town.

Crow intended to decorate the Anatole's atria and other parts of the structure with art he had collected in his travels during the past quarter of a century, but that collection would not suffice. The following year Crow and others attended a large antique auction in Los Angeles and elsewhere and made major purchases for the hotel. Other art-buying excursions followed.

From the first, Crow said he wanted the Anatole to be the nation's "premier brick job." He considered that while the in-vogue colors were tan and brown, red had a classic appeal, and even while in disuse in certain periods was always revived. Shelmire now began traveling once again, attempting to locate bricks of just the right color and texture for the project.

In addition to the brick, Crow intended to embed sculpture in the walls and include other objects of interest. He selected Mara Smith, an art professor at the University of North Texas, to design the bas relief sculptures and produce them in red clay before the art bricks were fired. Each brick was numbered and a retired journeyman bricklayer set them in their proper places. Crow maintained tight security at the site to ensure none of the bricks would be disturbed and so disrupt the sequence.

Crow also decided to include a 1,000-seat formal auditorium in the Anatole, for industrial shows and conventions. Plans for additional restaurants were soon added. All the while Crow also considered room and suite locations, china, glassware, caterers, and so forth. The job was more than just planning a hotel building; Crow had to create a small city, find people to staff it, and of course consider matters of landscaping, drainage, parking, and the like. Crow had always preferred to handle several deals simultaneously and had been able to do so with relative ease even when planning and executing large projects. This time it was different; the Anatole took more of his attention than any other undertaking in his career.

Cooper later said that as far as he knew nobody had ever done anything like that before. "Crow had more damn fun planning and designing the hotel than in any other project in the preceding 25 years." He added that if Crow had been on a desert island he would want to build the next sandpile better than the one before, and the Anatole was his best sandpile.

He would call me and he'd say, Bill, I think up on the parapet of that roof if the bricks were canted at 45 degrees wouldn't it make it a better parapet? I said Trammell, I don't give a damn. I don't think anybody's going to look up there. But he'd worry about that for three or four weeks and get the architects to draw it up several times, and finally they came up with what he wanted. He was involved in designing the interiors and the ballrooms and the restaurants—he had more fun than anybody you ever saw and of course, but he will generally do something like that.

The Anatole was inaugurated on December 15, 1978, and opened for business on January 1, 1979. It was a success almost from the start, unusual in that it ordinarily takes two or three years for a hotel like that to break even. Encouraged, Crow planned an addition. IBM owned the 25 adjoining acres, and George Shafer was able to work out a deal whereby Crow purchased land in another location and traded it for that holding. Now Crow executed his addition, a 27-story structure known as "the Tower." It too had an atrium, shops, restaurants, and larger ballrooms and more meeting rooms, along with 720 additional guest rooms, making Crow's gem a 1,620-room facility.

On this occasion his timing turned out to be poor; the Tower opened in 1981, in the midst of a recession, and was not profitable for several years. As far as Crow was concerned, however, it was an artistic success that in time would be a commercial one as well. He never regretted the lavish attention he bestowed upon it. In any case, a person who had risen to his estate might have looked upon this not only as an investment, but worthwhile therapy. That, at least, was Bill Cooper's judgment. Even while monitoring the Anatole construction, he wondered what might have happened if Crow had left the scene before it was completed. Cooper suggested he would have had to take over the project himself. "Had that happened," Cooper confessed, "it would have been the nicest Holiday Inn you ever saw."

NINE

New Beginnings

"Why are some of the partners leaving?"

The reasons they've expressed include size of the Company, their roles in the new Company, and personal lifestyle goals. The combination accelerated longstanding but somewhat latent, philosophical differences. But that's okay. And, done correctly, it is normal and proper to retire. We all will some day. We regret any partner's decision to withdraw and certainly will miss any partner who leaves. But in the past decade, 27 partners have left and nearly all have done so on good terms, being bought out or staying on deals depending on their desires and those of their replacements. We anticipate that out of these retirements some great opportunities will come to deserving partners and leasing agents. Upward mobility and meritocracy are alive and well.

—JOEL PETERSON, 1987

In 1972, when everything seemed to be going well, Crow made a small concession to the worriers at the Dallas headquarters and the field. He commissioned McKinsey & Co. to conduct a study of operations. The

McKinsey team swarmed over the headquarters, and quickly became confused by the unique arrangement there and in the field as well. After sorting matters out McKinsey concluded that the partners were "dedicated to as little change as possible," but recognized modifications were coming due to expansion of the business, Glaze's desire to retire, and Crow's mortality. The consultants came up with a series of recommendations, including:

- Greater interlocking of ownership on a selected basis
- Cross-business moves of a few (2–3) high-potential partners
- Discussions with all key partners on personal philosophies and goals
- Developing explicit strategies and plans for each entity to be financially self-sufficient while maintaining control and financial development.
- Increasing the Crow children's responsibility and independent accomplishments in the business

If Crow gave any consideration to the advice it was not reflected in operations. In any event, such matters were of minor importance during the Crunch, and the study was shelved and soon forgotten. However, the issues raised by McKinsey in 1972 reappeared several years later, and all were finally addressed by Crow, Williams, and Peterson.

The decade following the restructuring at Trammell Crow Company was a difficult one for real estate development, and indeed for many industries and occupations. This had little to do with events in Dallas, but rather decisions made by the nation's political leaders and world events.

While Crow, Glaze, Williams, Peterson, and Leventhal were completing their efforts at keeping the Crow interests intact, there was a new twist in the inflationary spiral, during which interest rates rose to levels not seen since the Civil War. Until the autumn of 1979 the Treasury and the Federal Reserve Board had struggled to keep rates down, largely by expanding the money supply, but such action only led to additional inflation. That this situation would be sharply altered became evident on October 6, when Federal Reserve Board Chairman Paul Volcker announced that thenceforth he would ignore interest rates and focus instead on the money supply. The meaning of this action was immediately apparent to economists, and soon after to builders and others involved in real estate as well; the central bank would abandon attempts to keep rates steady. In effect, Volcker proposed to end the inflationary spiral by permitting it to run its course, knowing that along the way there would be much economic pain but in the end, he hoped, steady prices and a more healthy economic environment.

Signaling this intention, the Fed raised the discount rate a full point, to 12 percent. On the next Monday many banks responded by hiking their prime rates to 14½ percent. Money creation slowed. Measured in terms of M 1 (currency plus checking accounts) the supply rose by an annual rate of 6.8 percent from December, 1980 to June, 1981, dropped to 5.9 percent for the next six months, and then, in the six months ended June, 1982, to 4.6 percent.

Interest rates advanced spectacularly in the months that followed. The federal funds rate (the rate banks charge one another for loans) went from an average of 7.9 percent in 1978 to 11.2 percent in 1979, then on to 13.4 percent in 1980, and there were times in the next two years when market rates rose over the 20-percent level. Home mortgages were available, but at rates of over 22 percent, causing grave distress at thrift institutions in the business of granting such loans.

Had the Crow partners foreseen the direction and magnitude of interest rates in the years following the Crunch they well might have abandoned all hopes for survival. As it turned out the market value of their properties, also affected by inflation, rose faster than ever. As they had in the mid-1970s, investors who had been burned by sell-offs in stocks and bonds turned to real estate as one of the few safe havens for their funds, in the process bidding up their prices and leading to overbuilding in many parts of the country. In this period, too, favorable treatment of real estate developments under the existing tax code made investments in this area most attractive to those who found themselves in higher tax brackets due to inflated incomes.

Major real estate investors who bought and held properties then rather than selling them usually paid little or no income tax, no matter how much their properties rose in value. It was no secret that over the long run carefully selected and monitored real estate was a superb investment. However, most considered the trouble involved and potential nonliquidity as major barriers. Given the situations in the securities markets in those years, along with tax considerations, they were impelled toward commercial real estate.

This pushed prices higher. Office buildings and shopping malls no longer were simply places to conduct business, but investment properties as well.

The housing market was also good in those years, as home seekers, seeing prices rising sharply, made purchases and accepted those high mortgages, thinking they might be even more exorbitant a few months later.

There was increased activity for the Crow partners. For a while it seemed business would continue to boom. 1979 was a busy year for Crow. It was then that he opened the Anatole. A 1.7-million-square-foot enlargement of the World Trade Center was also completed, along with a 500,000-square-foot addition to the Apparel Mart. Ground was broken for the Diamond Shamrock Tower in downtown Dallas, and plans were being drawn up for what would become the LTV Tower, which is today's Trammell Crow Center. This expansion in Dallas was reflected in most of the markets in which Trammell Crow Commercial functioned.

While noting that 1979 had been the best year for his varied interests, Crow realized it couldn't last. All the projects completed that year had been financed before the upswing in interest rates. Business was bound to be more difficult when the cost of money rose while the economy softened, and especially if the tax code were altered. It wasn't that Crow expected a return to hard times like those of the mid-1970s, but rather that, even while it was alien to his nature, he was preparing for a slowdown. "I don't know when Dallas will ever again grow at the rate it saw in 1979—if ever," he told a reporter in early January, 1980. It was not the kind of statement one would have expected from him prior to the Crunch, and a clear indicator of the alteration in mood at the company.

In the summer of 1979 Crow predicted that "beginning within the next few months or year, real estate activity, with the exception of home building, will decline drastically—30, 40, maybe 50 percent," adding that "I further believe that the decline will last for several years." To be positioned for this kind of business environment, Crow, Williams, and Peterson sent out the word to cut back on land purchases, instituted new economies in Dallas and the field, and considered additional alterations in the structures at Commercial and Residential.

The reasons for the transformation were many and complex, but in the end they boiled down to one: The economic and business climate of the early 1980s was drastically different from that of the late 1970s. The economy fell into a recession that turned out to be the worst since the Great Depression of the 1930s. The unemployment rate, which in 1979 had been 5.8 percent, went over 10 percent by autumn of 1982. By then, however, the Federal Reserve Board's medicine of permitting interest rates to rise in order to stifle inflation had clearly worked. The inflation rate was 4 percent, and the home mortgage rate had fallen into the single-digit range, stimulating construction. Lower rates were energizing the commercial market as well, but competition was keen and margins slender. The inflationary era that had begun in the late 1960s had ended, and the various real estate interests had to come to terms with this new business environment.

Confusing matters were singular developments on the investment scene. Unwilling to believe that the curb to inflation was more than a passing phase, convinced that prices would ratchet ahead within a year or so, investors demanded high yields from bonds and assured growth from stocks before committing funds to securities. Bond yields thus remained high, affording investors the highest real returns on loans in a century. In the summer of 1982 the stock market embarked on one of the most sensational upward sweeps in history, which would take it from Dow 770 to over 2,500 in six years. Just as investors had turned to real estate when stocks and bonds turned sour, now many switched back to securities and away from real estate.

Finally, there were problems arising from the long-anticipated changes in the Internal Revenue Code, which shook up the entire industry. Under the new regulations real estate investments lost several tax preferences, and the investment aspect had to be rethought.

There was a particularly damaging factor in Crow's situation. With the exception of New Jersey, relatively few of his operations were in the northeastern quadrant of the country. In the 1970s New England and the middle Atlantic states had been troubled, due in part to high energy costs, and for the same reasons the energy-producing areas of the nation prospered. In the early 1980s, however, oil and natural gas prices collapsed, and then it was the Southwest's turn to suffer. The economies of such states as Texas, Oklahoma, and Louisiana were in shambles. Houston was particularly hard hit, with Dallas not far behind. Since so many rigs and related equipment were repossessed, by the mid-1980s the back lots of some Texas banks and thrift institutions looked like oil equipment dealerships. The regional slump caused many Texas and Oklahoma banks to face severe crises. The outlook was decidedly cheerless in that part of the country.

Trammell Crow Commercial and Residential both came through this stagnant economic environment in good shape, vindicating many of the changes brought about in 1976 and 1977, and the companies experienced extraordinary growth in the years following the recession. In 1976 the various Crow interests had assets of approximately $1 billion. By 1982 the figure was close to $3 billion, and in late 1986 the Crow companies' combined assets were some $13 billion. The 15 offices of 1976 by then had expanded to 90, the 300 employees to 5,000.

The scope and thrust of the Crow interests had also expanded. There was increasing recognition of the Trammell Crow name, as for the first time the firm abandoned its traditional low profile. Sheer size presented a pleasing problem, but a problem nonetheless. As TCC expanded it became more important for Williams and Peterson to maintain resolute leadership, for the more competitive real estate markets of the mid-1980s presented new and far more complex trials as well as providing slimmer and not as automatic rewards as Crow had enjoyed in his early years.

The leadership had to be careful to maintain internal harmony at the same time. Increasingly Williams, Crow, and Peterson had to arbitrate jurisdictional squabbles between the regional and local partners. There were disputes between Commercial and Residential regarding boundaries of competition, and both entities had difficulties with the Crow family interests, which were also expanding at this time, occasionally into areas believed the domain of one of the other Crow companies. Residential partners quarrelled over territories. Additional companies were added to Crow Residential in the 1980s, so that by 1986 there were six of them, each eager for opportunities, with young partners eyeing one another's backyards, convinced they could perform better than those in the field if only given the chance to do so.

There also were related difficulties regarding compensation. In the mid-1980s the various Crow units were hiring around a hundred new leasing agents a year, more M.B.A.s in fact than most investment banks, and it also had become necessary to bring in senior people to fill major administrative posts. As was the tradition, the TCC regional and local partners received a half share of each deal in their territories, which was to be shared with the junior partners. As it became increasingly evident that promotion from project partnerships would become more difficult, Dallas wanted to create a formula for compensation that would provide greater rewards for junior partners.

Williams had ample reason to be deeply concerned about partner satisfaction. The country had been carved into regions, each with its own partner, and all the partners jostled against one another in struggles for territory. The days when Crow searched for local real estate agents and developers to represent his interests were long gone. Some of the newer partners looked askance at their seniors, feeling possibly that they were equally or better qualified, and would be in their positions if only they had arrived when the company was young. Given this situation, it perhaps was natural they would believe their shares of deals should be larger than what was allocated by the regional partner. Dissension arising from such concerns might easily poison the atmosphere and spread, if not checked in a way all viewed as fair.

Competition was keener than ever. Large financial institutions became more interested in real estate. Pension funds, trust funds, union welfare funds, and the like now entered the field directly, while the insurance companies, always involved, now were less content to be passive investors and sought shares in new enterprises. There were more major companies in the field. Paragon, Koll, Simon, Hines, and of course Lincoln presented challenges of a different magnitude than those faced in the 1960s and 1970s. Thirty years earlier Crow would drive to a local bank to obtain financing for a warehouse or two. By the mid-1980s it was an almost daily occurrence for one or more teams of managers of large funds to visit Crow headquarters to inquire about pending projects, unblinkingly offering to invest hundreds of millions of dollars at a clip. Representatives of foreign interests also wanted to know of properties to purchase. Crow became one of the leaders in selling to foreign investors. They would listen, take notes, and then go on to other developers to learn what they had to offer.

It made sense for Crow to enter the investment real estate field in a more deliberate and organized way than previously had been the case, to act as a broker-dealer as well as a principal. Moreover, Trammell Crow's original axiom, that the buyers of commercial real estate had been right and the sellers wrong, had to be rethought. Why retain properties if the market in a particular part of the country appeared to be weakening? Of course the Crow interests had much experience in disposing of properties, having been obliged to do so during the Crunch, but this time it was different. Beginning in the early 1980s TCC embarked on a regular program of identifying and then selling certain properties from its portfolio, in some cases to increase liquidity and in other to reshape the portfolio as markets evolved. A logical next step was to purchase properties that seemed undervalued and then sell when the market for them improved.

By 1985 these buying and selling operations had evolved to the point where it made sense to gather them together in a new division, to be known as Trammell Crow Ventures (TCV), which handled acquisitions, dispositions, and asset management. TCV was headed by David Clossey, formerly the managing partner for the Dallas office of the national law firm of Jones, Day, Reavis and Pogue. In 1987 Ray Golden, senior partner and chief financial officer at the Manhattan-based investment banking firm of Salomon Brothers, joined Crow. Golden intensified Crow's efforts in investment banking. It was no minor business; in 1987 TCV's annual acquisitions were more than $1 billion and dispositions were running at the rate of $500 million. In addition, TCV developed pools of capital for Crow, including $200 million for a new real estate investment trust, and put together Trammell Crow Equity Partners as a pool of cash from outside investors, and Trammell Crow International Partners, a fund for Japanese institutions.

The changes in the tax code opened new windows of opportunity for TCV. Because of the elimination of several deliberate benefits to developers, capital costs on many projects exceeded anticipated earnings. Financings were becoming more complex. Many banks and savings and loans, especially in the Southwest, were troubled, and their shrinkage obliged the Crow partners to consider going directly to the capital markets for their financial requirements. Far better terms could be obtained from well known and financially strong national concerns than from smaller, local ones.

It was also becoming apparent that high-quality financial and accounting standards would be required if profits were to be maximized. The Crow interests had to consider the desirability of having in-house capabilities in some areas in which they traditionally used outside vendors, such as investment banking. Already financings were being made through the issuance of commercial paper, which resulted in annual savings running in the millions of dollars. And if the decision to expand such activities and enter related fields were made, would it then make sense to attempt to market expertise and knowledge to other firms? There were no clear answers to such questions at the time, but in 1985 the advisory services and capital markets operations were added to Ventures.

Thus Trammell Crow Ventures became a significant force at Crow, leading both TCRC and TCC into new deals and approaches to the business. For example, by 1988 TCV had entered into an arrangement with Merrill Lynch whereby that giant investment bank would place certain Trammell Crow residential properties with its institutional clients. Increasingly the company was becoming a force in American finance as well as real estate. Indeed, the two were becoming inextricably intertwined. What had started out as a small operation concerned with industrial real estate had branched out into commercial real estate, and then into housing, marts, hotels, and other real estate ventures. Now the firm was rapidly becoming what might be described as a real estate–based financial services complex.

One reason for the evolution was a momentous change in attitudes toward financial statistics. Ray Golden observed that while in the past companies were most concerned with their profit and loss statements, in an age of takeovers they were devoting more attention to their balance sheet. Most industrial companies tend to mismanage—or even fail to manage—their real estate holdings. Appreciating this, corporate raiders take over companies, liquidate part of the real estate portfolio, and use the proceeds to repay loans made to undertake the venture. Williams and Peterson believed that better management of real estate could make companies less susceptible to takeovers, and in 1988 these views were receiving an increasingly respectful hearing in corporate boardrooms. For example, TCV was engaged by the Santa Fe Southern Pacific Railroad to review that company's real estate assets.

The difficulties of savings and loan institutions and banks in the Southwest provided yet another area of opportunity for TCV. Many of these institutions were burdened by properties repossessed when loans went into default; they had large sums of money tied up in properties that they had no interest in holding and managing. At the same time they suffered from liquidity problems, which obliged the Federal Home Loan Bank and the Federal Savings and Loan Insurance Corporation to enter the picture.

In some respects the dilemmas facing the banks and savings and loan institutions resembled the difficulties the Crow interests had faced in the Crunch. Few organizations were as well prepared to handle such problems. When they could apply marketing and redevelopment expertise, the Crow partners acquired properties from these institutions. In other instances they purchased properties they thought would improve in price during the anticipated economic upturn. These dealings with institutions prompted the creation of a TCV subsidiary, Trammell Crow Advisory Services, which managed assets for the Federal Home Loan Bank and for the Federal Savings and Loan Insurance Corporation. The largest contract was for the management of Western Federal, a major, failed Dallas-based savings and loan that had $2 billion worth of assets. Crow sent in a team of 20 people that developed a workout for the Western Federal assets on a fee basis.

Another area of opportunity resulted in the creation of the TCV subsidiary, the Capital Markets Group, which operated as an investment bank, assisting financial managers with respect to funding individual transactions. Capital Markets representatives made calls on financial institutions to try to develop lasting and profitable relationships. They focused on areas that were new for the Crow companies. For example, Capital Markets devoted a large amount of time in Japan attempting to position the Crow operating companies with major Japanese lending institutions. In addition, Capital Markets was successful in obtaining management contracts for several deals that in the past might have been taken by Wall Street banks. Such operations were one of the reasons Ray Golden had come to the firm—to provide investment banking expertise. Soon after Golden obtained representation in the key New York market.

The Advisory Company and Capital Markets, and Advisory Services were logical outgrowths of existing expertise, but the move into acquisitions and dispositions troubled some of the partners who were perfectly content with the established rules and disliked the generation of new units that might intrude upon their operations. They were even more troubled by a development in the real estate operations. National corporations were approaching developers asking for national service and terms, and some of Crow's competitors were already providing both. In order to compete effectively on this front Crow would have to do the same. For this reason TCC launched a National Marketing program, through which Dallas would seek out opportunities from companies having national interests. All these moves were not only quite different from anything Crow had done when erecting warehouses in Dallas in the 1950s, but even far removed from operations during the late 1970s.

The Crow interests were changing rapidly, being transformed into a new kind of company that even those at the vortex of change had trouble comprehending fully. Change is always disquieting, even when welcomed, and not all in the company hailed these new moves. Partners in the field were naturally deal-oriented, and their deals were usually local. Now they saw the Crow interests taking on a national flavor, and they weren't at all certain this evolution was in their interests.

How far might the transformation lead? Expansion into foreign markets merited investigation and was one diversification in which Crow had experience. Unfortunately, as has been seen, much of the experience had been bad. Crow entered these markets when America's world role was starting to decline. American business had entered the world in force in the postwar period, but soon the world would discover the American market. By the late 1980s American developers were busily marketing properties to European and Japanese investors rather than the other way around, and Crow had a large stake in this business, which was far more important than his activities abroad.

Today Trammell Crow Company operates in only a few markets overseas, but international business is not high on the company's agenda, accounting for only 2 percent of assets. One reason is the perception of even better opportunities in the domestic arena. There also remain scars from the earlier encounters. Even so, a global economy called for a global company. Under the circumstances, it was to have been expected that various parts of the company would examine their approaches to foreign markets.

While concerned with managing and directing change, the Management Board, including Crow, had to deal with the delicate matter of continuity. This hadn't been much of a problem while Crow was the center of all activity, directly involved in all deals. Even in the early 1970s there had been projects Crow knew about but had never seen. Now there were many more of them, and many more partners whom he, Williams, and Peterson could not monitor, as once had been possible. Greater surveillance and supervision was felt necessary. "We needed to manage the Crow reputation better in the marketplace," said Williams, "And we needed to put a leadership structure in place for the next decade."

To accomplish this, in 1981 Williams recommended the creation of the Trammell Crow Foundation, a structure which would act to hold the Crow family ownership interests and accumulate capital. In return for leaving capital in the firm and the right to use the name, the new foundation would receive a 99-year right to a percentage of all the company's deals. In this way, the organization took on yet another aspect of a corporate identity: immortality, the ability of an enterprise to survive its founders and operators. Thus the once highly unstructured series of deals was becoming increasingly regularized, if not centralized. It seemed that too would happen eventually.

The deals did become regularized, but gradually and only after much reflection. Modifications were accomplished through a series of steps, each one debated, discussed, revamped, and tested by those involved. It was a slow, often torturous process, and at times the partners entertained differing notions of where they wanted to head. But the mechanism was set in motion, and once underway, an inner logic seemed to be leading it on.

The process began in 1982. As the real estate industry started to expand again, Crow, Williams, and Peterson initiated discussions about how to meet challenges posed by several factors: increased size, territorial considerations, compensation, the need for more and synchronized services, the creation of new businesses, the existence of more intense competition, and the rapidly changing tax codes.

The new status of the Crow family businesses was also of concern. Soon these would take on the title of Trammell Crow Interests (TCI), headed by Harlan Crow. TCI did not include the Market Center, which remained under the management of Lucy Crow Billingsley. The new entity did include a variety of enterprises that did not fit into either the Commercial or Residential company. Among these were hotels, medical buildings, TCDC, the 75,000 acres of farmland owned in Louisiana, Arkansas, and Mississippi, the 3.5 million square feet of marts and office space in Europe, and several miscellaneous operations.

The varied and seemingly constantly multiplying Crow entities would have to become better coordinated than before, if only to avoid chaos and turmoil. Coordination implied even more controls from Dallas. More national, as opposed to local and regional, planning had to be undertaken, to capitalize on reputation and assets. The local companies would remain, of course, but under the umbrella of a national concern. The goal was to unify all the Trammell Crow pursuits, bringing together formerly disparate and occasionally fractious elements.

For example, it was becoming apparent that while the 1977 reforms had gone far toward providing Commercial with a sense of structure, one was still lacking at Residential. Dallas wanted to impose a structure, but this would require the assent and cooperation of a group of possibly recalcitrant partners who had a major stake in retaining the old relationships. Still, everything seemed to point in the direction of further centralization, more controls, more paperwork, and the generation and dissemination of far more information, along with a term heard more often at the firm, and with derision: "the bureaucracy."

Added controls and restraints were bound to meet with opposition from some of the partners, who not only were concerned with their earning levels but enjoyed the large degree of autonomy that existed. Some still grumbled about restrictions placed on their activities in the wake of the Crunch, and they could be counted on to raise objections to any new programs that might hinder their actions. A few members of the Management Board itself were critical of the direction the leadership proposed the company should take. Several partners stated their case and, seeing their views would not prevail, decided to retire or in some other way leave the company, while others engaged in the debate continued to voice their opposition during the years that followed, and managed to alter programs and plans.

Complicating the situation was the simple passage of time. Those hungry, young partners of the early 1970s had become older and wealthier. In some cases, just as Crow had feared, they had become more cautious and less aggressive. Departures were more common as time passed; partners retired, went off on their own, or simply cut back on commitments. Of the 23 partners in the Crow companies in 1973, only seven remained in 1985. More new partners were being named all the time; the path to partnership was more open than ever. Even so, those scores of young leasing agents attending annual meetings looked carefully at one another, knowing competition was keen.

One of the most pressing tasks facing the veterans was training the wave of newcomers, impressing upon them the nature and importance of the Trammell Crow tradition, and making certain they put into practice the principles Crow had developed, which brought the various interests to their successful status. Orientation could not be successfully accomplished unless there was some agreement as to just what the heritage meant, and which aspects were to be stressed. Achieving consensus, then, was also on the agenda in the mid-1980s. Through all the changes, Williams and Peterson conducted a campaign of education and persuasion worthy of political candidates in national elections attempting to appeal to local interests while offering a more expansive vision as well.

The evolving nature of the economy and their industries left the various Crow interests little choice but to change their focus and methods. For example, by mid-decade competition had become more intense. The commercial side of the business was greatly overbuilt because of tax considerations, the deregulation of the thrifts, and an overly optimistic view of future needs. Some of the gloomier prognosticators were saying that office buildings in some cities could perform in the second half of the 1980s the way the oil drilling industry did in the first half. Already some skyscrapers in Houston were under the hammer, and vacancy rates of 20 percent were not unusual in Dallas.

By the mid-1980s, with inflation hardly a problem, tenants would not sit still for what in the past had amounted to almost automatic rent increases, and in some cities they were demanding—and getting—cuts and concessions of a kind not seen since the Depression. In such an environment the limitation of risk was a matter of high priority. The risk could often be diluted by taking on more outside partners, especially for costly ventures. By 1985 the several Crow companies had approximately 50 investors/lenders, typically pension funds and insurance companies. This was yet another way the business had changed, and another way Dallas might serve the local and regional partners.

By the mid-1980s the local partners were being swamped by problems brought on by changes in the economy, zoning and regulations, growth and antigrowth attitudes, competition, and the ever-more-complicated atmosphere in which they operated. Once Trammell Crow himself could handle all the details regarding design of a warehouse, land acquisition, the hiring of architects and builders, and just about anything else that came along. Over time he had to take on assistants, but by the 1980s even these could not handle the intricate and complicated issues the partners were facing. Knowledge and expertise were required in a wide-ranging number of areas, which was why in the 1970s the Crow interests had established the Professional Services Group, which was geared to serve TCC itself, but also provide services to the partners on a fee basis.

Professional Services is made up of four units: Controller, Legal, Tax, and Administrative, all headquartered in Dallas. The Controller Group personnel soon became involved in a probing study of cost controls and a search for additional ways to maximize income. One investigation revealed that the TCC companies had consistently undercharged for leasing commissions and in the area of management and developer fees; rectifying this situation could lead to income running in the tens of millions of dollars annually. Leasing agents often concentrated on obtaining the initial tenants for warehouses and office buildings and slackened their efforts when it came to renting the last 15 percent or so of space. Since the final tenants often meant the difference between profits and losses, clearly a greater effort would be required on this front. In fact, management of properties appeared to have always been less than efficient, but this deficiency was particularly damaging in the difficult markets of the 1980s. Some partners billed late for services, in effect financing their tenants; in 1985 an internal audit revealed over $2.5 million in unbilled services, with some of the collections almost a year late. The same audit uncovered the fact that while TCC obtained beneficial contracts from janitorial and related companies, little of that came down to the bottom line. One of the Dallas partners, Greg Young, estimated that if Crow charged tenants the same rates as did smaller developers and landlords who did not receive the benefit of reduced charges, "we would increase cash flow by half a million dollars in Dallas [alone]." Such audits saved the partners an estimated $10 million between 1984 and 1987.

The Legal Group offered the partners specialized expertise at fees substantially lower than those charged by top law firms. It soon attracted a small group of talented attorneys led by Kathy Smally and Tom Green, both of whom had clerked for U.S. Supreme Court Justices and had then gone on to become partners in important law firms. The Legal Group possessed skills not found in most law firms, due to a knowledge of Trammell Crow's particular businesses. In addition, Legal handled partner buyouts, a key aspect of its work.

The many and complicated alterations in the real estate tax code, and the Tax Group's complete concentration on such matters, made its services attractive to the partners. Typically real estate people are asset rich but illiquid. Large tax payments could require substantial property dispositions, which is why partners were eager to cut the tax burden by using the expertise of the Tax Group.

Finally, the Administrative Group dealt with personnel recruiting for field offices as well as for Dallas, and also could handle payrolls, brochure preparation, food services, and a wide variety of other matters.

The sharing of expertise had advantages, but some partners shied from some of these services, preferring to obtain them locally, and they remained wary of Professional Services as yet another sign of centralized control from Dallas. Cherishing their sense of independence, the deal-driven partners in the past had chafed at paperwork and report making, and avoided at least some of this work. If profits weren't as high as desired, they argued, additional deals made possible by having the extra time for working on them more than compensated for inefficiencies in keeping track of details. But Dallas was able to document evidence that centralization was no nickel-and-dime matter, that savings could be substantial.

The issue came home dramatically in 1985 when Peterson reported that over the past 10 years Dallas had raised, documented, and serviced

$450 million in equity at rates running from 10 to 20 percent lower than the competition, made possible by utilizing the Crow name and placing large-scale loans. The total incremental savings for this one set of operations provided the Crow organizations with a savings of from $45 to $90 million. Centralized tax preparation effected savings of approximately $2 million a year. Coordination of syndication fees brought savings of $4 million over 3 years, while sophisticated cash management, made possible through economies of scale, generated $1.6 million in income annually. In 1985 three construction audits disclosed overpayments of $800,000, which were recovered.

Dallas was able to illustrate just how the lack of centralized controls and harmonized efforts hurt the company. For example, partners occasionally found themselves contending against one another for significant projects. Crow recruiters from different parts of the country bid against one another for newly minted M.B.A.s, who afterward might attempt to switch from one unit to another in order to realize better terms.

Centralized efficiency had its price: In 1976 the costs of centralized services of all kinds were $1.4 million; by 1981, when Dallas initiated new services, the amount was slightly more than $3 million. The addition of National Marketing, new computer systems, enhanced accounting and internal auditing, as well as the move to new quarters in the LTV Building, caused such charges to rise above $17 million by 1985. If some partners complained about the growth of such charges, Peterson could observe that the savings and benefits derived from them amply justified the expenses. "Over the past ten years of the restructured Trammell Crow Company," he wrote in 1985, "we've had nearly $1.5 billion in equity growth for in-house partners."

Not all the changes proved worthwhile. Attempts to centralize data processing did not bring the kinds of benefits once thought possible, and changing technologies made the concept less desirable, so it was soon abandoned.

The changes were having a big impact on morale. Each partner looked upon his territory as sacrosanct. The new Trammell Crow enterprises might decide to operate in the partners' bailiwicks, a situation they found intolerable.

In 1986 came the long-awaited attempt to create a more unified structure for Residential. The matter was becoming more pressing. Not only were partnership disputes increasing, but the business climate was clouded.

Some of the problems derived from alterations in the tax laws, such as elimination of tax-free bond financing and the lack of tax equity under the new legislation. In addition, as with TCC, there was a generalized overbuilding and softness in several of Residential's most important markets.

Peterson had taken command there on a temporary basis, his charge being to fashion the kind of organization that would harmonize activities and bring an end to partner conflicts, while at the same time maintaining partner satisfaction and cooperation.

The initial step was to define the business. Peterson suggested it might be well to consider that while TCC served the real estate needs of businesses, what he hoped would become the more-structured TCRC would meet the real estate needs of individuals. This was no mere exercise in semantics. Dallas was carefully drawing the line between TCC and TCRC. The definition of the divisions' primary businesses was to be the litmus for deciding any future problems regarding turf.

TCRC's primary ventures were rental apartments, community development, and for-sale housing, all areas in which it already was well represented. The secondary markets were hotels and several businesses that were fairly new for the companies—hospitals, retirement homes, club management, and mobile home communities. One indication that Residential intended to expand into new fields could be seen in suggestions that Crow might create and manage vacation resorts. Here too care was taken to ensure that boundaries between the Crow businesses were clear. "Any potential involvements in the hotel or medical development businesses must have Crow Family approval."

The restructuring thus began with simple matters of definitions and the more difficult one of establishing an objective. The next step was to create a tactical program that the Residential partners could accept, and then to implement it.

An agenda was prepared and presented in the summer of 1986. There would be a total of 20 "steps" in reshaping operations, beginning immediately with the designation of roles to be played by key persons, a definition of exactly what partnership entailed, and the creation of an allocation formula for partners. It was planned that by January, 1987 a management board would be in place, and resolution would be made regarding the ongoing relationships with the Crow Foundation (meaning the family) interests. Williams, Crow, and Peterson hoped to have administrative procedures in place by the end of 1987 for what by then would be the restructured companies.

The senior partners had initiated formal deliberations on the future relationship between TCRC and TCC. That there were continuing animosities and differing work styles between Commercial and Residential was obvious. Peterson spoke of a "veiled antipathy" between the two units. "Some at TCC view TCRC as an aggravation, a pretender to the Trammell Crow name, a later-arrival," while "many at TCRC see TCC as competent, big, better-known and decidedly more structured." Peterson added that "some have found TCC partners uncooperative." For their part TCRC partners felt they had nothing to apologize for, given their recent performance—in 1986, on a smaller employee and asset base, it had a larger cash flow and almost as many starts as TCC, as the accompanying table illustrates.

RELATIVE PERFORMANCE OF TCC AND TCRC IN 1986

	TCC	*TCRC*
Employees	2,200	750+
Total Assets	$8 billion	$1.2 billion
In-house Equity	$1.7 billion	$100 million
Recurring Cash Flow	$37 million	$38 million
New Starts	$1.7 billion	$1.5 billion

Source: Trammell Crow Company

Each company remained wary of the other. TCC was coming to see a more highly structured TCRC as a potential competitor; already some of the TCRC partners were talking about expanding into mini-warehouses and other structures long deemed the province of the Commercial company. But it cut both ways; the slack commercial market in some parts of the country caused some of the TCC partners to consider residential development.

The partners were quite different. TCC was dominated by leasing and marketing people, who made up 40 percent of the force, while 35 percent of TCRC people were in construction and finance and another 20 percent in property management. Construction and financial people could become partners at TCRC, while almost all the TCC partners were in development. The tone at TCRC was more breezy and informal, and quite a few of the partners had come in as partners, with substantial outside experience. In contrast, the TCC partners were quieter, wore dark suits, and preferred to grow their own new partners.

The Trammell Crow Interests division carried great weight in the discussions. Not only were Trammell's opinions respected and sought, but the family was far and away the most important owner of both companies, with $500 million net worth in TCC and an ongoing 15- to 17½-percent interest in new business through the Foundation, while it had some $60 million in TCRC and a 27-percent interest in future developments. Crow believed in the need for bridges between TCC and TCRC, but also saw the necessity for a wall. TCC should stay out of residential construction, he said, and he refused to permit TCRC partners the use of his name for commercial development. Crow supported centralization and greater cooperation, but at the same time understood the importance of maintaining separate structures at TCC and TCRC. They were different businesses, with different kinds of customers, and different knowledge and expertise requirements. Having gone through the experience himself, Trammell had concluded that no one could run both businesses simultaneously.

So the two entities would be separate but coordinated. By early autumn of 1986, the senior partners were considering the means whereby this could be done. There probably would have to be an overseeing board that would include partners from both companies. Closer coordination in matters of support systems seemed quite possible. And in time it might be feasible to select single partners to oversee both businesses in specific parts of the country. This possibility was a troublesome matter, since in some instances enmity existed between the local Residential and Commercial partners.

There was a continuing need to convince partners in the field that the new structure would not threaten their well-being or impinge meaningfully on their independence. Despite the centralization that had taken place, the senior partners retained much autonomy and power. Some might be inclined to go off on their own if this independence was seriously threatened. It had happened before; Trammell Crow not only made his partners millionaires, but out of the Crow organization had come more than a dozen large, independent real estate companies. One of the major problems in the restructuring would be to make certain there was no significant hemorrhaging of talent.

The problem first materialized in late summer, when one of the Management Board members telephoned regional partners to suggest a meeting of the operating partners in Dallas, without Crow, Williams, and Peterson, to examine several questions, paramount among them being the allocation of percentages of deals and concern their shares would be diminished. They also wanted to consider the leadership of the company, some elements of its administration, and the relationship of the Interests to the other parts of the company, all of which were sensitive subjects. Central to these issues were the questions of how much control Dallas ought to have and the freedom and rights of the regional partners. Should Dallas set policy when the business was in the field? It was to be an open exchange, with the partners informing Crow, Williams, and Peterson of their intentions and the agenda beforehand. The meeting did take place, and the accumulated fears and doubts of several years erupted.

Williams and Crow were dismayed and Crow was incensed that they would have this kind of assembly. Crow spoke with several of the partners who had called for the meeting. They asked him not to be offended, adding that they thought there were some matters that had to be ironed out, and they couldn't speak freely with him, Williams, and Peterson in the room. Afterward there were conferences with these partners. At the end it appeared that calm had returned, and that aside from some ruffled feathers and bruised egos, all would go well. Even so, the tensions remained.

The debate was completed by the end of September, by which time Crow, Williams, Peterson, and Harlan Crow embarked on a series of meetings with all the partners to discuss the changes to be implemented starting on January 1, 1987.

What they called the "new Trammell Crow Company" would consist of four operating companies under a management board comprised of Crow, Williams, Peterson, Teague, and the national partner of each of the operating companies, with Clossey as chief financial officer. To head TC-Commercial, Crow and Williams required a person who was experienced, trusted, and respected by all in the firm, and they selected Bob Kresko. Bailey, Childress, Simmons, and Shafer were designated as national operating partners. By now there were 14 regional partners instead of the previous 4, and 3 of them—Bailey in Southern California, Bob Speicher in Florida, and Robert Watson in Denver—were also partners in TC-Residential in their parts of the country, bringing about the beginning of a more unified company. Terwilliger remained at the head of TC-Residential, which had 7 regional partners with an opening for a partner for elderly housing and several other special programs. TC-Interests, operated by Harlan Crow, now had five discrete units: Public Distribution, Hotels, Medical, Foreign Operations, and a catch-all including farms and other specialized areas. TC-Ventures, headed by Clossey, remained as before.

The "unified" Trammell Crow Company generated both success and problems. A new group of regional partners provided fresh energy and insights in leading local operations, and the flow of information improved. Regular meetings of all the TCRC partners resulted in a wider sharing of information; partners in New England received leads from their counterparts in Florida, and news regarding sources of financings was shared regularly. On Harbor Island in Florida Bob Speicher coordinated efforts among TCRC, TCC, and TCV in creating mixed-purpose developments. The company was learning how to react creatively to new challenges. For example, there were skyscrapers with commercial development on the lower floors and residences higher up. In the past, had either TCC or TCRC attempted to bid on a mixed project like this, the other would have protested. Now they cooperated on such ventures. At first TCC would not use TCRC construction partners on their projects; this started to change, and the cooperation seemed certain to expand—slowly, to be sure—in the future. To the suggestion that the company might still fly apart when Crow passed from the scene, one partner replied, "That might have been true five years ago. Not now."

Even so, the new dispensation resulted in departures. Perhaps it was inevitable that some of the partners would reject so sweeping a change, and of course all of them had the resources and knowledge to make a go of it on their own. As it was, 15 percent of the partners left the firm, to retire or go off on their own.

The first departure occurred in Chicago, where Allan Hamilton expressed dissatisfaction with the plans being made. He had been dispatched to that city in 1968 to take charge of commercial development, with a territory that eventually extended to Minneapolis-St. Paul, Detroit, and Milwaukee, as well as rights to Canada should the Crow interests ever enter that market. At the time of his arrival Crow had a dozen or so warehouses in the region. After 18 years in the section Hamilton had put up more than 50 structures, including a complex of 2 million square feet in office buildings and a hotel Crow insisted be called The Hamilton. Hamilton had suffered during the late 1970s, when the Midwest in general was in such trouble that it was known as the "Rust Bowl." At that time he was assisted by Dallas. In the mid-1980s, however, the situation reversed, as Hamilton's territory was doing well while Texas suffered from the oil glut.

Hamilton was strongly opposed to National Marketing, which he perceived as an unwarranted intrusion on his territory. In his view, National could come in and make a deal to accomodate a client who was important in some other part of the country, even though the local partner considered it a poor use of resources. Hamilton also had a rift with Williams; a plan of his to start work on a 1-million-square-foot building on speculation, without long-term financing, was vetoed, one of the few times in TCC history that a local partner was not backed by Dallas. His suspicions that Dallas would attempt to take control were not unlike the fears felt by other Crow partners.

For Hamilton the breaking point came when he was passed over for promotion to a national partner and two new regional partners, Jon Hammes and Kirt Woodhouse, were named out of his region. As a result, Minneapolis-St. Paul, Milwaukee, and Detroit were allotted to those partners, leaving Hamilton with the Chicago area and the Canadian market, where the company had fledgling operations in Calgary and Edmonton. Hamilton flew to Dallas in January, 1987, met with Crow and Williams, and demanded the right to purchase the entire Crow commercial holding in his region. When this proposal was rejected, Hamilton and seven of his local partners walked out of the office. Soon after they organized a competitive operation, Hamilton Partners. In the end the properties were divided. "The Dallas partners changed things so drastically that I could not live with it," said Hamilton afterward, adding, "It's a lot like a divorce." There were suits and countersuits, which unfortunately ended with hard feelings and recriminations.

There may have been factors other than the restructuring involved in Hamilton's departure. In some respects he was typical of those Crow partners who, upon achieving considerable financial success, wanted both independence and the opportunity to pursue other interests. Hamilton owned a ranching operation in Montana and a working farm in Illinois, to which he devoted a considerable amount of time. His departure was painful, but it may have been the best way out for all interested. Hamilton went off on his own, opening the way for several young partners eager for the opportunity to prove themselves.

His leaving set a bad precedent, however. Not since Gil Thomas's separation had a partner left with such rancor. It appeared certain there would be more departures, and because of the experience with Hamilton Dallas was concerned that they would be litigious, bitter, and in general leave a sour and demoralized feeling at headquarters and in the field.

Hamilton's departure was not incomprehensible, but its abruptness and hostility were unexpected. The severance of Ned Spieker, the regional TCC partner in the Northwest, was in some ways different and in others similar. Ever since talk of restructuring and greater centralization had started, Spieker had indicated his dissatisfaction with the concept and plan. This was not something new; Spieker was one of those who, after achieving the heights at and with Trammell Crow, yearned to be off on his own.

What troubled Spieker most at this point, however, was what he percieved as the lack of a clearly defined sense of territoriality and a deepening disagreement with most of what Williams and Peterson were attempting in the way of unity within the Crow operations. Even in the 1970s he had complained when Lincoln entered what he considered his area and bid for office buildings and warehouses. Crow was still an owner at Lincoln at the time, so in effect he was competing with his own regional partner. Spieker protested to Crow, who while not condoning the practice didn't stop it either. Pretty soon, Spieker recalled, "It got very tense."

Spieker held little hope for the Williams approach. "You know what Don Williams did. We have this tattered old goose, and it just limped along. It didn't look good. It wasn't explainable. But it kept laying golden eggs. And then, Don wanted to streamline the goose. Make it explainable. Make it a professional-looking goose. And then the eggs just stopped."

In Spieker's view, the strength and viability of the Crow operations rested on the maximum independence of the partners, and not in further coordination of efforts. Williams wanted to create an entity in which partners would have a great deal of autonomy, but stand prepared to accept necessary coordination from Dallas; Spieker was unhappy with regulation. He thought the tree should be pruned regularly—that is, he expected some healthy "self-trimming" of partners leaving the business. He intended to be one of these. Also, like Hamilton, Spieker had other interests. Having amassed considerable net worth, he wanted time for other pursuits, both in business and recreation. He had invested in a California winery and a tortilla bakery, as well as several other businesses, to which he had devoted an increasing amount of time and attention. As was the case in Chicago, Spieker's departure opened opportunities for several of the young partners—these from the West Coast—who now moved up a notch.

Perhaps Spieker would have exited under any circumstances. As it was, in March, 1986, together with 3 other TCC partners in the region, Spieker informed Dallas of his intentions to leave the firm. "The reorganization highlighted some philosphical differences between us and the company," he later explained. "We wanted a little more free-whelling, personal type of atmosphere." Spieker instituted the process of dividing properties soon after Hamilton broke away, organizing with what had become 10 other Crow partners in California and Oregon a new firm, Spieker Partners. Compared to the break with Hamilton, this separation went smoothly, with a minimum of rancor, and with residual friendship.

Spieker's departure left a vacuum in one of the nation's most important markets. As it happened, for several years Joel Peterson had wanted to relocate to California, where his wife had roots. Now he took the opportunity to do so, taking charge of TCC's Northwest regional business from new offices outside of San Francisco.

This change came on the heels of a "swing around the circuit" that Peterson undertook in February, 1987. He had met with partners and brought the message to them from Dallas, answered questions, and in general calmed apprehensions. Most important, he went to listen to complaints and concerns and then responded to them. By the time he returned to Dallas, Peterson was able to identify four major areas of concern. These involved opportunities for new partners, leadership of the new Trammell Crow Company and means of identifying future leaders, the matter of size, and just how the Crow family would function in the revamped operation.

Peterson responded to these issues against the backdrop of Hamilton's departure and the workout of separation with Spieker. The corporate culture of Trammell Crow Company was undoubtedly changing. What Peterson had to do was indicate to the troubled partners that much of the old would remain, that the company ran no danger of becoming bureaucratized, and that their futures were ensured. His remarks to concerned partners as he traveled the Crow circuit often began with the supposition that the old status quo had not been so favorable to the partners after all. Peterson tried to recall for them how it used to be.

Before the combination, only one new region had been formed in 10 years, over two-thirds of Trammell Crow Partners was owned by 12 people, the Crow

Foundation had a non-dilutable position, and in certain regions new partners were coming in at 5-8 percent interests. In addition, no new Firm Partners had been elected in over two years and the recommendation of the Strategic Planning Committee to significantly increase partner salaries had been on the table for nearly a year.

The status quo had been changed in ways the partners could applaud. In the past few months, Peterson pointed out, nine new regional partners had been appointed, new cities were colonized, the Crow family had agreed to dilute its share in deals over time, and there would be a review of salary policies. Significantly, "almost no single partner's economics were reduced anywhere."

Next, Peterson moved to calm apprehensions about leadership. Firm Partners would have a say in who would serve as Firm Managing Partners, ensuring those in the field participation in central planning. In this way a meritocracy would exist in Dallas as well as around the country. Williams and Peterson would continue to arbitrate differences, work with the Crow family's interests, and in general represent the Trammell Crow Company in its overall business stance. Peterson pointedly observed that "no one from the field seemed anxious to help administer Firm business," reflecting the continued aversion to managerial tasks in the developer-centered atmosphere that had always existed at TCC and TCRC and their antecedents. This underscored Williams's and Peterson's contention that, however grudgingly, most partners conceded that a more secure structure was needed. However, all wanted to work deals and to remain as far from administrative tasks as they could get.

As to complaints about the expansion of administration in terms of amplitude and authority, Peterson noted that Dallas was continually striving to reduce overhead and streamline operations, and remained on guard against unnecessary bureaucratization. During the past few months the central staff had been reduced from 250 to 150 and the 12 department heads had been consolidated into 4 positions. Accounting services had been farmed out, while computer systems development had been turned back to the regions. In all, more than $1 million in Dallas overhead had been shaved or shifted elsewhere. Regional overhead accounted for 80 percent of all overhead costs, and Peterson hoped to lower Dallas's portion even more.

Perhaps the most pressing matter in the minds of the partners was the Crow family's relationship to the new Trammell Crow Company. Peterson reported that thoughtful consideration had been given to the matter, with several changes being made and new mechanisms created. In the first place, the Interests' 17½-percent share in projects would be cut, eventually being reduced to 11.7 percent. Its voting power in TCC would be reduced from 33 percent to 24 percent, and the Foundation would be restructured under Harlan Crow, who was clearly to assume his father's role as head of that branch. The Family would increase its liquidity, including the raising of over $500 million through the refinancing of the Market Center. Peterson knew that his responses could not allay all the partners' anxieties, and that a continuing dialogue would be necessary during the transitional period.

By early spring the company looked somewhat different than it had on New Year's Day. Kresko remained as Managing Partner for Commercial, with responsibilities for supervising four national partners. Tom Simmons headed operations in the Northeast, Don Childress had the Southeast, while Gary Shafer's domain was a wide swath in the middle of the country running from the Great Lakes to Texas. Terwilliger continued as Kresko's counterpart at Residential. One of his charges was the difficult task of designating regional partners, and he situated Bob Spiecher and Dick Michaux as national partners.

The Management Board was enlarged to include new partners for all operations. The board indicated that its composition was only a preliminary table of organization, presented as much for discussion as anything else. Everyone realized the situation was extremely fluid.

And it would always be so. In May, 1988, Williams informed the partners that Don Childress would be leaving to form his own company. Ten years earlier the Trammell Crow Company's fledgling Managing Board was made up of Williams, Crow, Shutt, Hamilton, Spieker, Shafer, Kresko, Simmons, Brown, and Myers. Half of these men were no longer with the company. That partners would depart seemed quite clear, as was the fact that the reassigning of partners would continue to be a crucial task. The concept of the evergreen company was very applicable. When a poorly performing partner withdrew, it was no great loss, while the company had learned too that when a highly productive partner resigned, it was akin to a child going off on his or her own. Trammell Crow Company prized partners who were original and innovative, but people with those qualities are not often willing to subordinate themselves to any great extent to a structure they do not control. At one time, in the 1960s, Trammell Crow was known as the man who made millionaires of his partners. This was still true in the late 1980s. Moreover, he also had a hand in training the leaders of some of the industry's more important real estate development firms, who had literally gone to school under Trammell Crow, and it would appear that this too would continue to be the case. As Joel Ehrenkranz put it, "Trammell Crow is a large tree that throws off many acorns."

Whatever disruptions were caused by the nature of the Hamilton, Spieker, and Childress departures were mitigated in part by the next withdrawal. In June, barely a month after the Childress announcement, Kresko told the 1988 convention of TCC partners and marketing representatives that he intended to step down as Managing Partner the following month. Kresko would remain on as Regional Partner "so long as I am needed," but he hoped to leave by his 55th birthday to pursue private interests.

For years the company had struggled to find a paradigm for a proper, clean, and mutually satisfactory means whereby partners could withdraw harmoniously, without rancor or bitterness. There had been some cases of harmonious departures, such as Mack Pogue's break in 1977. At that time many of the future leasing agents were in elementary or junior high schools. More than 40 percent of the agents in 1988 were attending their first TCC annual meeting; there were few grey heads in the crowd, and anyone over the age of 35 would have felt somewhat ancient. Only a handful of partners at the firm could recall the circumstances of the Crow-Pogue split. The new Trammell Crow Company did not so much lack a history or tradition as it did a collective memory that might serve as a stabilizer, a glue to hold the partners and agents together. What is not known cannot be remembered, and a majority of those at the meeting could not have known how Pogue withdrew—but they had witnessed the parting of Hamilton, Spieker, and Childress.

It may have been that part of the collective memory was created that day. Kresko explained that he and his family were quite well off financially thanks to his activities with Trammell Crow, and that any additional moneys he might have earned would only make some charities and other recipients of grants richer than they might have been. This might have been his view, but the almost 500 partners and leasing agents in the audience may have heard and seen something else. Some of these recently recruited young people were uncertain what to expect. They were eager to learn and to please, and possessed the kind of hunger and excitement Crow himself had always sought in his trainees. Now they were treated to a graceful withdrawal that may become the norm in the future. They also heard a speech from the person who was to succeed Kresko as Managing Partner, Peterson, who just as he had done for the past half dozen years moved to fill a gap in the structure, subordinating his private interests to the good of the whole. Also, Williams announced that Bob Whitman, already a national partner in Trammell Crow Investments, would become chief financial partner and, along with Teague, a national partner in the firm. In addition to hearing about changes in personnel, the attendees were able to share information and experiences with one another in what had to be an important socializing event.

At that meeting Trammell Crow remarked that some time around 1978 he realized that he did not know all of the leasing agents. He hadn't seen all of his buildings, let alone getting to know all his young people personally. How could he, given the size of the enterprise?

This raised an issue that few at the company care to discuss. Being large, more unified, and national was not an unalloyed blessing, or for that matter in the eyes of some a blessing at all. The Trammell Crow interests were dedicated to serving all parts of the country, and they also wanted eager young people in every market, seeking out customers and deals. As the company expanded, some might have asked if this was a wise strategy. Might it not make sense to avoid certain markets due to their unpromising nature? And if agents were assigned to difficult markets, what should the company's attitude be toward rewarding performance? In 1988, for example, Baton Rouge and Denver were weak markets, while San Diego and Chicago were strong. A partner or leasing agent in the former might have 1 deal or rental for every 10 turned in by someone in the latter. The success rate did not mean the San Diego personnel were superior to those in Denver; in fact, just the opposite might have been the case. And if this indeed were so, how should compensation and rewards be made? How might the Denver people be encouraged, and the San Diego counterparts be kept on their toes? As the decade drew to a close such questions were not answered. That they would have to be addressed seemed manifest, if only to maintain the morale of those young leasing agents upon whom the future of the enterprise depends.

Whether or not Peterson and Terwilliger will be able to inculcate in these young people a sense of community to go along with their ambitions is of vital concern, as is the matter of continuity. Characteristically real estate companies have been vehicles for ambitions and aspirations of individuals. Unless they are family concerns, and small ones at that, they tend to liquidate with the disappearance of the founder from the scene. In the 1980s the Trammell Crow Company attempted to fashion a structure that would outlive Crow, Williams, and whoever follows.

One of the enduring clichés of business is that no company or organization is stronger than its people. But the strength comes from many facets of the people's attitude and vision. Acceptance of a common tradition and code of conduct is important. A willingness to embrace the concept of teamwork while retaining individualism is another. A vision of a common future certainly is a *sine qua non* for any organization. The attitude does not spring from the pages of archives and training manuals; *people* have to accept, hold, personify, and then pass on such views to their successors. If Crow and Williams are to succeed, the Crow people will have to see themselves as part of a continuum, who not only will do well for themselves but for future generations as well.

In 1988 Don Williams spoke of the kind of young people he hoped to attract to the interests, repeating the firm's stress on brains, character, and experience as Trammell Crow and he had done so often.

We still are after a combination of characteristics in a person. One is intelligence, you know, the old business of smart folks do better than dumb folks. Secondly, they have a willingness to work hard, and a willingness to take less now to build for the future. So you get less current pay with us, less than any other place that our trainees could get a job. That was true in the '60s, and that's true with us today. So, you're betting on the future. And thirdly, is the high personal standards, high ethical standards: character.

So what I've told our people is this: We're looking for two things in our business: One is brains, and the second is character. If somebody has brains and character, we can give them experience if they work hard. If they don't have brains and character, we can't give them that.

"Brains and character" were determining factors from the start. Recall the typical experience of Tom Simmons in 1970, when as a young man with no experience he was thrown into the negotiations for 2001 Bryan, one of Crow's major projects, which at other firms would have required the talents of several experienced developers.

I remember when he took me into a meeting with the contractors and all the lenders and introduced me to them. He built me up as though I was something that I really wasn't, which I think demonstrated confidence in my abilities. I couldn't believe that he would give someone as young and as inexperienced that kind of responsibility. I think this has been pretty characteristic of his career. In many cases it has paid off. He's brought good people on quicker than they ever would have come with other companies. In some cases it has backfired, but on balance it has been a worthwhile and successful policy.

While the formulation may sound simple enough, it really isn't. There are many with character, and perhaps more with brains. Experience in real estate, as in any other field, can be accumulated. How many are there with *both* brains and character, along with a willingness to work hard, to accumulate and learn from experiences, and to sacrifice rewards in the present in return for greater ones in the future? That set of characteristics pretty much describes Crow himself in 1948, when he met the Stemmons brothers, peered into the Trinity Basin, and decided to build a warehouse there.

BIBLIOGRAPHY

Books

American Guide Series. *Texas: A Guide to the Lone Star State.* Austin, TX: Texas Highway Commission, 1940.

Bainbridge, John. *The Super-Americans.* New York: Holt, Rinehart and Winston, 1972.

Dillon, David. *Dallas Architecture, 1936–1986.* Dallas, TX: Texas Monthly Press, 1986.

Faulk, Odie B. *Texas after Spindletop, 1901–1965.* Austin, TX: Texas State Historical Association, 1952.

Fehrenbach, T. R. *Lone Star: A History of Texas and the Texans.* New York: Macmillan, 1968.

Frantz, Joe B. *Texas: A Bicentennial History.* New York: Norton, 1976.

Garreau, Joel. *The Nine Nations of North America.* Boston: Houghton-Mifflin, 1981.

Meinig, D. W. *Imperial Texas: An Interpretive Essay in Cultural Geography.* Austin, TX: University of Texas Press, 1969.

Oppenheimer, Evelyn, and Bill Porterfield, eds. *The Book of Dallas.* Garden City, NY: Doubleday, 1976.

Richardson, Rupert. *Texas: The Lone Star State.* Englewood Cliffs, NJ: Prentice-Hall, 1970.

Sumner, Alan. *Dallasights: An Anthology of Architecture and Open Spaces.* Dallas, TX: American Institute of Architects, 1978.

Interviews

Anderson, Jacob	July 2, 1980
Bailey, Tom	January 13, 1988
Baldwin, Peter	July 21, 1980
Beck, Henry	August 22, 1980
Berry, Harold	June 16, 1980
Biddle, Jim	August 28, 1980
Billingsley, Lucy	April 7, 1981
Bowles, Donald	June 24, 1980
Box, Cloyce	August 13, 1980; January 12, 1987
Brannon, L. Travis	April 1, 1980
Bromberg, Henri	April 4, 1980
Bronkema, Jim	July 6, 1980
Butcher, Preston	May 7, 1981
Campbell, Bill	July 7, 1980
Carter, Frank	November 3, 1980
Caswell, James	April 1, 1981
Clossey, David	January 4, 1988
Coker, Jim	September 29, 1980
Collins, Barbara	August 12, 1980; June 16, 1987
Collins, Jim	January 8, 1981
Colwell, Winston	December 26, 1980
Cooper, Bill	May 16, 1980; May 20, 1980; January 13, 1987; January 8, 1988
Cox, John	September 30, 1980
Crow, Davis	August 24, 1980
Crow, Harlan	October 2, 1980; January 6, 1988
Crow, Howard	August 13, 1981; August 17, 1981
Crow, Kathleen	August 22, 1981
Crow, Margaret	October 21, 1980; January 13, 1987
Crow, Michael	April 21, 1981; June 15, 1987
Crow, Robert	May 8, 1981
Crow, Stuart	August 17, 1981; August 25, 1981
Crow, Trammell	January 12, 1987; January 5–6, 1988
Crow, Trammell S.	January 7, 1988
Cullum, Earl	October 23, 1980
Dawson, Kim	June 25, 1980
Denis, Read	May 7, 1981
Dillard, Bill	June 30, 1980; January 13, 1987
Dona, Tony	January 7, 1988

Dubinsky, Gus	May 17, 1980
Echols, Daws	July 23, 1980
Fleck, Lawrence	January 27, 1981
Francis, J. D.	August 14, 1980
Francis, Raymond	August 18, 1980
Gilliland, Burton	August 5, 1980
Glaze, Robert	September 30, 1980; December 19, 1980; June 15, 1987
Golden, Terry	April 16, 1981
Haggar, Ed	June 25, 1980
Halford, Lee	June 16, 1980
Hamilton, Allan	December 10, 1980; February 25, 1988
Hamilton, Hope	June 17, 1987
Hancock, J. D.	November 20, 1980
Harrison, James	August 28, 1980
Harrison, W. A.	August 7, 1980
Hay, Jess	February 24, 1981
Hodges, Sam	April 1, 1980
Holbrook, Charles	April 23, 1981
Hood, Robin	May 5, 1981
Hudson, Jim	June 26, 1980
Huffines, Leonard	July 17, 1980
Hunt, Pancho	July 22, 1980
Jackson, Bill	August 21, 1980
Jacobson, Jake	January 14, 1987
Johnson, Hubert	January 23, 1981
Jones, R. B.	January 29, 1981
Jordan, Grady	June 24, 1980
Justice, Glenn	August 19, 1980
Kahn, Edmund	May 23, 1980
Kresko, Robert	December 10, 1980; October 23, 1987
Kristal, Gary	February 26, 1988
Lansburgh, Robert	August 6, 1981
Ledyard, Gibby	June 15, 1980
Leventhal, Ken	April 15, 1988
Linde, O. J.	May 6, 1981
Long, Virginia	August 13, 1981
Lowe, Jack	August 12, 1980
McNeff, George	September 9, 1980
Melville, Mary Lou	November 7, 1980
Middleton, Bob	April 27, 1981

Miller, E. D.	September 4, 1980
Miller, Henry	July 9, 1980
Moran, Jim	July 21, 1980
Myers, Mark	April 14, 1981
Peterson, Joel	April 14, 1981; January 12, 1988
Phelps, Robert	July 1, 1980
Philipson, Herman	August 14, 1980
Pogue, Mack	January 5, 1981; January 13, 1987
Pope, Ewell	April 7, 1981
Presley, Dewey	September 4, 1980
Rogers, Mazie	August 1, 1980
Rogers, Ralph	November 12, 1980
Schafer, Gary	June 15, 1987; January 6, 1988
Schoppe, Bill	April 26, 1981
Shutt, Tom	October 7, 1980; June 18, 1987; January 8, 1988; February 26, 1988
Simmons, Tom	December 10, 1980
Simon, Heinz	June 18, 1980
Smally, Kathy	January 6, 1988
Spieker, Ned	December 10, 1980; April 15, 1988
Stalcup, Joe	September 25, 1980
Stemmons, John	May 21, 1980; January 5, 1988
Stewart, Robert	September 15, 1980
Stone, Kenneth	March 3, 1981
Strauss, Ted	September 17, 1980
Sullivan, Roger	June 17, 1980
Talkington, Perry	July 24, 1980
Teague, Tom	April 21, 1981; June 16, 1987
Thompson, Wade	August 27, 1980
Wallace, Denny	September 10, 1980; January 14, 1987
Watson, Howell	June 16, 1980
Wellman, Emory	April 27, 1981; February 26, 1988
Wilhite, Clifton	September 16, 1980
Williams, Don	February 25, 1981; March 4, 1981; April 9, 1981; May 11, 1988
Wolfe, Frank	September 5, 1980
Wood, George	October 5, 1980

NAME INDEX

SUBJECT INDEX

(Buildings are in Dallas unless otherwise noted.)

Standing, left to right:
Trammell S. Crow, Trammell Crow, Wade Crow, Shirley Crow, Trammell Billingsley, Stuart Crow, Carter Crow, Harlan Crow, Bob Crow, Howard Crow, and Henry Billingsley.

Seated, left to right:
Barbara Crow (with baby Daniel Crow), Margaret Crow, Lucy Billingsley (with daughter Anne Billingsley), and Paige Billingsley.

On floor, left to right:
Katy Crow, Margaret Crow, Nathaniel Crow, and George Billingsley.